INTRODUCTION TO THE
THEORY OF LINEAR SYSTEMS

Introduction to the Theory of Linear Systems

E. A. FAULKNER
M.A., Ph.D., C.Eng., M.I.E.E., F.I.E.R.E.
Senior Lecturer in Physics
University of Reading

CHAPMAN AND HALL LTD
11 NEW FETTER LANE LONDON EC4

Distributed in the U.S.A.
by Barnes & Noble, Inc.

AUTHOR'S PREFACE

This book was written primarily for students of physics and of applied physics, with the aim of providing them with some insight into the way in which linear circuits are approached by the modern electronics engineer; also of helping them to understand the general properties of linear susceptibility functions, particularly the Kramers-Kronig relations. The author hopes that the book may also be useful to the trained electronics engineer as a brief and informal summary of results which are discussed in much more detail in specialized texts.

In developing the logical framework of the subject, a special effort has been made to bring out the essential unity of Laplace-transform methods with Fourier-transform methods, and to show how the transforms arise directly from physical considerations and not merely as a method of solving differential equations. For this reason, the bilateral Laplace transform has been used throughout, rather than the unilateral transform usually employed in engineering texts. This approach does not lead to any difficulty in relation to further reading, provided that the student realizes that the 'unilateral Laplace transform' of a function $f(t)$ is essentially the bilateral transform of the function $u(t)f(t)$, where $u(t)$ is a unit step function at $t = 0$.

For detailed discussion of the more elementary aspects of the subject, the student is referred to

C. M. Close: *The Analysis of Linear Circuits* (Harcourt, Brace & World 1966)

and for more advanced reading to

A. Papoulis: *The Fourier Integral and its Applications* (McGraw Hill 1962)

S. J. Mason & H. J. Zimmermann: *Electronic Circuits, Signals, and Systems* (Wiley 1960).

These works contain further references which do not need to be repeated here.

In the writing of this book, the author has been greatly assisted by discussions with Professor C. W. McCombie and Mr R. S. Leigh, and by comments on the text from Professor P. B. Fellgett and Mr P. Atkinson. Invaluable help with the preparation of the manuscript has been given by Mr M. L. Meade.

E.A.F.

CONTENTS

Author's Preface   *page* vii

1   Introduction   1
1.1   Systems: input and response   1
1.2   Linearity   3
1.3   Practical limitations   5
1.4   Electrical analogues   6

**2   Impulse response, convolution integral, and the system
      function**   7
2.1   Singularity functions   7
2.2   Impulse response   11
2.3   The convolution integral   15
2.4   Step response   16
2.5   Complex exponential input   18
2.6   The system differential equation   19

3   Use of Fourier and Laplace transforms   24
3.1   Fourier series and Fourier transform   24
3.2   Laplace transform   28
3.3   The inversion integral   31
3.4   Second-order system   38

4   Frequency-response functions   41
4.1   Application of the Fourier transform   41
4.2   Fourier transform of a causal function   43
4.3   All-pass, low-pass, and symmetrical band-pass systems 44
4.4   Group and signal-front delay   51
4.5   Minimum-phase criterion   54

5   Integral theorems   57

5.1 Integration of $(s{-}p)^n$ 57
5.2 Initial-value theorem   58
5.3 Real-part integral theorem   59
5.4 Integral of $[H(j\omega)]^2$ 61
5.5 The Kramers-Kronig relations   62
5.6 Relationship between gain and phase   64
5.7 Poles and zeros within a contour   67

6   Negative feedback   70

6.1 The concept of negative feedback   70
6.2 The feedback equation   71
6.3 Non-linear distortion   74
6.4 Second-order feedback system   76
6.5 Nyquist criterion   78
6.6 The analogue computer   82

Index   87

CHAPTER 1

INTRODUCTION

1.1 Systems: input and response

In this book we are concerned with physical systems which involve time-varying measurable quantities, $x(t)$ and $y(t)$, between which there is a causal connection. We visualize an experimental situation in which x can be varied at will and the corresponding variation in y can be observed. In such a case the variable x constitutes the *input* to the system, and y is the *response* (often called the *output*). This relationship may be represented by equation (1.1.1)

$$x(t) \rightarrow y(t)$$
$$\text{input} \quad \text{response.}$$

$$(1.1.1)$$

Here the arrow denotes a causal connection.

We shall be dealing only with systems whose characteristics do not vary with time. Let us suppose that some particular input function $x_1(t)$ gives rise to an output function $y_1(t)$, and let us consider the output that we should have obtained if we had delayed the input function by a fixed time τ. If the system is *time-invariant*, this delayed input function $x_1(t-\tau)$ would give rise to an output $y_1(t-\tau)$, whatever the value of τ. This means that in writing down the input function for a time-invariant system we are free to define the time origin in any way that we please, and when we are dealing with input functions which are bounded in time we shall usually find it convenient to define the instant $t = 0$ as the instant when the input function was 'switched on'.

As an alternative to the formulation of equation (1.1.1) we may represent the system mathematically as an *operator* $\mathscr{S}$:

$$y(t) = \mathscr{S}x(t).$$

$$(1.1.2)$$

The operator $\mathscr{S}$ can be regarded as a set of instructions which enables us to transform the whole of the function $x(t)$ into another

1

function $y(t)$. If x and y have the same dimensions (e.g. if they are both voltages or both linear displacements) then the operator $\mathscr{S}$ is dimensionless. On the other hand, if x and y have different dimensions, the units of $\mathscr{S}$ must be specified.

Since the instantaneous value of a single physical quantity is represented by a real number, $x(t)$ and $y(t)$ are both real functions; thus the operator $\mathscr{S}$ must have the property of transforming any real function of t into another real function of t.

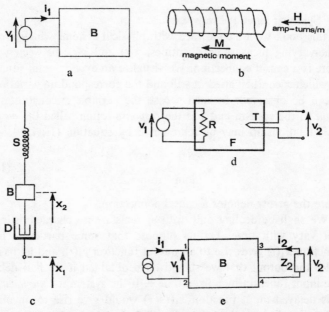

Figure 1.1 (a–e). Some examples of systems.

In figure 1.1 we show examples of some simple systems with their inputs and responses. In figure 1.1a the 'black box' B represents an electrical network of some kind, the input being the voltage $v_1(t)$ and the response being the current $i_1(t)$. Similarly, in the magnetic system shown in figure 1.1b the input is the applied magnetizing field $H(t)$, and the response is the magnetic moment $M(t)$.

Figure 1.1c shows a mechanical system consisting of a bob B attached to a coil spring S and also to a dashpot D. (A dashpot is a component which embodies the principle of viscous friction in that it exerts a force proportional to the rate of change of its length). The input is $x_1(t)$, the displacement of the lower end of the dashpot, and the output is $x_2(t)$, the displacement of the bob.

In figure 1.1d a furnace F is heated by a resistance element R driven by a voltage generator $v_1(t)$, and the temperature of the furnace is measured by a thermocouple T which gives an output voltage $v_2(t)$. v_1 and v_2 may be regarded as the input and response respectively of this thermo-electric system.

As a further example, figure 1.1e shows a four-terminal 'black box' B with input current generator $i_1(t)$ connected to terminals 1 and 2, and a *load impedance* Z_2 connected to terminals 3 and 4. In this system we may distinguish between the *driving-point* relationship between input $i_1(t)$ and response $v_1(t)$, and the *transfer* relationships between input $i_1(t)$ and response $i_2(t)$ or $v_2(t)$. Terminals 3 and 4 are usually called the *output terminals* as distinguished from the *input terminals* 1 and 2 to which the input generator is connected. If B contains no internal energy sources, it follows from the law of conservation of energy that the power delivered to Z_2 cannot exceed the power obtained from the input generator. On the other hand, B may contain batteries (or the equivalent) and *active components* such as transistors or tunnel diodes and be capable of delivering more power into Z_2 than it receives from the input generator. Depending upon its principal function, the 'black box' B may be designated in various ways – e.g. as an amplifier, a filter, an attenuator or a limiter – but nowadays there is a great deal of overlap between such designations and, to give just one example, it may be equally correct to describe a given system as a selective amplifier or as an active filter.

1.2 Linearity

When we say that a system is *linear* we are using the word in a very specialized sense. We do *not* in general mean to say that the relation between a response $y(t)$ and an input $x(t)$ is of the form

$$y(t) = Kx(t)$$

with K a constant independent of t, although such a relation would describe one particular type of linear system. Generally, a linear system is defined as one which obeys the *principle of superposition*, which is itself defined in the following way. Suppose that an input $x_1(t)$ causes an output $y_1(t)$, and that an input $x_2(t)$ causes an output

$y_2(t)$; then an input $x_1(t)+x_2(t)$ will give rise to an output $y_1(t)+y_2(t)$. In operator terminology this statement is written

$$\mathscr{S}[x_1(t)+x_2(t)] = \mathscr{S}x_1(t)+\mathscr{S}x_2(t) \tag{1.2.1}$$

and if equation (1.2.1) is satisfied for all functions $x_1(t)$ and $x_2(t)$, then $\mathscr{S}$ is said to be a *linear operator*.

It is often convenient to state the principle of superposition in the less general form

$$\mathscr{S}[nx_1(t)] = n\mathscr{S}x_1(t) \tag{1.2.2}$$

where n is any number.

Before we can be sure of the physical meaning of the statement that a system is linear, we must establish three principles which are really implicit in it; these are definition of zero, stability, and causality.

(a) *Definition of zero.* In order to give meaning to the cause and effect association between the y's and the x's, we must define the origins of x and y in such a way that if the value of x were to remain zero for all time then the value of y would also be zero.

(b) *Stability.* If we are to suppose that it is possible to perform an experiment which will establish whether a system is linear, we must assume that the input $x_1(t)$ is bounded in time, so that at some value of t we can say that $x_1(t)$ has ceased to be applied. It must also be true that the corresponding output $y_1(t)$ is bounded in time, so that at some value of t we can say that the system has ceased to respond to $x_1(t)$ and has returned to its original condition where $y = 0$. A system is said to be *stable* if it has the property of 'forgetting' about an input and returning to its quiescent state of its own accord.

In contrast to this we have the case of an *unstable* system where the response continues to increase for an indefinite time after the input has ceased, often in the form of an oscillation of increasing amplitude. Since in a practical system the response variable will not be able to reach an infinite value, the increase will cease when some physical limitation is reached and the system goes into a new state with different parameters. The linear system in its original form has now ceased to exist. Even in the absence of an input, the lifetime of an unstable system is limited because random fluctuations (such as thermal noise) in the input variable will provide an input which, however small, will eventually result in the destruction of the system in its original form. In discussing the theory of linear systems we are primarily interested in stable systems; however the properties of hypothetical unstable systems are of great importance, if only because they enable the conditions for stability to be established.

(c) *Causality*. The whole discussion has depended on the concept of a *causal* relationship between the output and the input. Now our general experience of actual physical systems tells us that the effect cannot precede the cause. Another way of saying this is that although we can easily construct systems which 'remember' the past, no actual system can be expected to 'predict' the future. A number of very important conclusions are drawn from the fact that the response $y(t)$ of an actual physical system must be zero for all time before the beginning of the input $x(t)$.

1.3 Practical limitations

The definition we have given of the linear system is in fact a mathematical idealization. In the case of an actual system, we very often have some physical reason for knowing that it approaches ideal linearity in the limit as the input amplitude approaches zero. Let us suppose that we are examining the linearity of such a system by successively applying inputs of $x_1(t)$, $2x_1(t)$, $3x_1(t)$ .... and observing the resulting response. It may be that the response to the second input is, within experimental limits, equal to exactly double the response to the first input and this fact establishes the system as a linear one over the range of input levels from 0 to $2x_1(t)$. However, this linear relationship will not continue indefinitely and, when the input exceeds some given value, we will begin to notice a departure from linearity which will become progressively more pronounced as the input level is further increased.

In many systems we observe the phenomenon of *saturation* where the magnitude of the response is more or less abruptly restricted to a limiting value. When a system has been constructed with the intention of its being a linear system, the departure from linearity with increasing input level is called *non-linear distortion*. In evaluating certain types of electronic systems we often find that a non-linear distortion level of the order of 0·01 per cent is very significant and may have to be measured with considerable accuracy. It is interesting to notice that the property of causality has quite a different philosophical status from that of linearity, and we never find it necessary to do careful measurements to determine whether our system is capable of predicting the future.

1.4 Electrical analogues

Because of the ease with which a linear electrical circuit with any required type of response can be both represented in a circuit diagram and realized in practice, linear systems of all kinds are frequently treated by analogy with appropriate electrical circuits. For this reason, and also of course because of its enormously wide field of application, the linear electrical circuit has a fair claim to be regarded as the archetype of linear systems and the discussion in this book will be almost exclusively in terms of electrical circuits.

CHAPTER 2

IMPULSE RESPONSE, CONVOLUTION INTEGRAL, AND THE SYSTEM FUNCTION

2.1 Singularity functions

If we are to analyse the behaviour of linear systems in a general way we must often make use of functions of a type known as *singularity functions*. Here we shall only be concerned with $u(t)$ and $\delta(t)$, the *unit step* and the *unit impulse* functions respectively. The latter is identical with the *Dirac delta function*.

The unit step $u(t)$, illustrated in figure 2.1a, is defined by the following properties

$$u(t) = 0 \text{ for } t < 0$$
$$u(t) = 1 \text{ for } t > 0 . \tag{2.1.1}$$

The value of $u(t)$ is thus undefined for $t = 0$. When considering its value in the region of $t = 0$ it is often convenient to use the symbols $0+$ and $0-$ to refer to values of t which are respectively an infinitesimal time after and an infinitesimal time before the precise instant $t = 0$. We can thus describe the behaviour of $u(t)$ at the origin by the following equations

$$u(0-) = 0, \ u(0+) = 1 . \tag{2.1.2}$$

In figure 2.1b there is shown a unit step which is not at the origin but at time $t = \tau$. This is clearly the same function as that of figure 2.1a apart from a change in the origin of t and we denote this 'shifted' function by $u(t-\tau)$. We can include this function in the defining equations by rewriting them in the form

$$u(t-\tau) = 0 \text{ for } t < \tau$$
$$u(t-\tau) = 1 \text{ for } t > \tau . \tag{2.1.3}$$

7

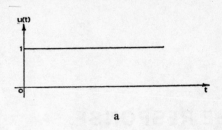

a

IMPULSE RESPONSE
CONVOLUTION INTEGRAL
AND THE SYSTEM FUNCTION

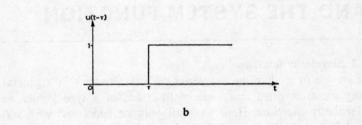

b

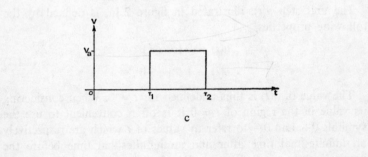

c

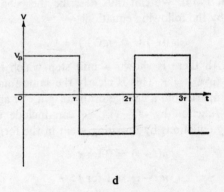

d

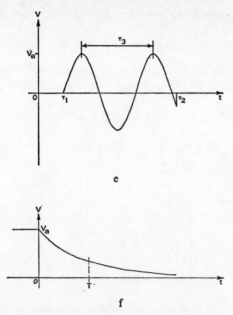

e

f

Figure 2.1 (a–f). The unit step function and examples of its application.

By the use of these equations it is possible to build up any 'rectangular' function of time, remembering that if we are to represent the time variation of a physical quantity then it will be necessary to use a multiplier having appropriate dimensions. The 'rectangular pulse' of amplitude v_a volts illustrated in figure 2.1c is thus described by the function

$$v(t) = v_a[u(t-\tau_1)-u(t-\tau_2)] \tag{2.1.4}$$

and the 'square-wave' of figure 2.1d by the function

$$v(t) = v_a[u(t)-2u(t-\tau)+2u(t-2\tau)\ldots] . \tag{2.1.5}$$

$= 2u(t-3r)$

The unit step function also finds application when we wish to formulate a function of time which is 'switched-on' at a given instant. An example of this is the switched sinusoid shown in figure 2.1e which can be represented by the product of a rectangular pulse and a sinusoidal function. It therefore has the equation

$$v(t) = v_a (\sin 2\pi t/\tau_3) [u(t-\tau_1)-u(t-\tau_2)] . \tag{2.1.6}$$

$\sin(2\pi[t-\tau_1]/\tau_3)$

Another example is the capacitor-discharge function of figure 2.1f, given by

$$v(t) = v_a[1-(1-\exp(-t/T))\,u(t)].\qquad(2.1.7)$$

We now turn to the unit impulse function $\delta(t)$ which is defined as the derivative of the unit step $u(t)$. Its properties are given by the relations

$$\delta(t) = 0 \text{ for } t \neq 0$$

$$\int_{-\infty}^{\infty} \delta(t)\mathrm{d}t = 1.\qquad(2.1.8)$$

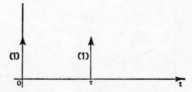

Figure 2.2. The unit impulse function.

The unit impulse is obviously not so easy to represent graphically, being a pulse of unit area, zero width and infinite height. In figure 2.2 the function $\delta(t)$ is represented by an arrow at $t = 0$, pointing towards ∞ and marked with the figure 1 to indicate the property of unit area. The impulse at $t = \tau$, also shown in figure 2.2, is the derivative of $u(t-\tau)$ and is the function $\delta(t-\tau)$.

An important feature of the impulse function is its so-called *sampling* property,

$$\int_{-\infty}^{\infty} f(t)\delta(t-\tau)\mathrm{d}t = f(\tau).\qquad(2.1.9)$$

Equation (2.1.9) is sometimes used to define the impulse function.

In order to represent an impulse of say, voltage, we must multiply the function $\delta(t)$, which by definition has dimensions of inverse time, by a quantity having the dimensions voltage–time. These are the same as the dimensions of magnetic flux, often represented by the symbol Φ, and we accordingly write the expression

$$v_1(t) = \Phi_1\delta(t)\qquad(2.1.10)$$

to represent a voltage impulse with area Φ_1 occurring at time $t = 0$.

Similarly a current impulse at $t = 0$ is represented by the expression

$$i_1(t) = q_1\delta(t) \tag{2.1.11}$$

where q_1 has the dimensions current-time or electric charge.

We shall see later that the impulse function has Fourier components spread uniformly over the frequency spectrum from zero to infinite frequency.

2.2 Impulse response

Let us now imagine an idealized linear system which is completely characterized by the operator $\mathscr{S}$. We are going to consider how the system would respond if we were able to apply an input in the form $x_1\delta(t)$ where x_1 is a parameter with appropriate dimensions as discussed in the last section.

The response to this input would be $\mathscr{S}[x_1\delta(t)]$ and we may define a *normalized impulse response* $h(t)$ (usually referred to, in cases where no ambiguity can occur, simply as the *impulse response*) by taking the ratio of this response to the area of the impulse which caused it, thus

$$h(t) = (1/x_1)\mathscr{S}[x_1\delta(t)] . \tag{2.2.1}$$

We note that if the input impulse had been centred at $t = \tau$ the corresponding response would have been $h(t - \tau)$. As we shall see in the next section, the operator $\mathscr{S}$ is completely characterized by its impulse response so that the response to *any input* can be predicted once $h(t)$ is known.

Having introduced the ideal case, we must now try to apply these considerations to an actual physical system. One obvious difficulty is that we cannot possibly make any physical quantity vary with time exactly according to $\delta(t)$ as defined in equations (2.1.8). Even if we do try to approximate to an ideal impulse (by using an extremely large value of the input quantity for an extremely short time) we are likely to find that we have exceeded the limits within which our system can be regarded as linear and that the system has gone into saturation.

The solution to this difficulty is to recognize the important fact that the impulse response of a linear system can be treated both theoretically and experimentally as a limiting case. *In evaluating the impulse response of a linear system we may consider the response* *to an input consisting of a rectangular pulse of width Δt centred on $t = 0$. In the limit as $\Delta t \to 0$, the normalized response to this input will be $h(t)$.*

This statement brings out the fact that every linear system has associated with it one or more characteristic times, usually referred to as *time-constants*. When an input pulse is much shorter than any of the time-constants the system 'sees' it as being instantaneous.

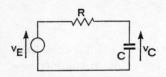

Figure 2.3a. Series *R–C* circuit.

We shall now consider the impulse response of a simple system. In figure 2.3a the input is a voltage $v_E(t)$ and the output is the voltage $v_C(t)$ across the capacitor. If $v_E(t)$ is a rectangular pulse of height v_1 the current i_C is initially given by v_1/R; if the pulse lasts long enough, the voltage across C will be an appreciable fraction of v_1 so that i_C will begin to fall. Clearly the 'short pulse' limit is reached when the pulse width Δt is so short that $v_C(t)$ is always negligible compared with v_1. This condition is

$$\left.\begin{array}{c} v_1 \Delta t / RC \ll v_1 \\[2mm] \text{or}\ \ \Delta t \ll RC . \end{array}\right\} \tag{2.2.2}$$

Provided that these equations are satisfied the charge imparted to the capacitor by the input pulse will be independent of v_1 and Δt separately, being dependent only on the area $\Phi_1 = v_1 \Delta t$. The only difference between the response to this finite pulse and the response to an ideal impulse of the same area is the finite slope of the response in the region of the origin due to the finite time occupied by the charging of the capacitor. If we now reduce Δt to such a small value that the increase in v_C in the region of $t = 0$ is 'instantaneous' within experimental limits, we will have reached a situation in which the input pulse is indistinguishable in its effect from a true impulse. In this extreme case we are justified in writing

$$v_C(0+) = \Phi_1/RC . \tag{2.2.3}$$

After the input voltage has fallen to zero, the capacitor discharges through the generator with a time-constant RC. In the limit as $\Delta t \to 0$, the discharge may be described by the equation

$$v_C(t) = (\Phi_1/RC) \exp{(-t/RC)} \ \text{for}\ t > 0 . \tag{2.2.4}$$

Since the capacitor was uncharged before the occurrence of the input pulse we also have

$$v_C(t) = 0 \text{ for } t < 0. \tag{2.2.5}$$

Now we may use the unit-step function $u(t)$ which enables us to combine equations (2.2.3), (2.2.4) and (2.2.5) into the single result

$$v_C(t) = u(t) (\Phi_1/RC) \exp (-t/RC) \tag{2.2.6}$$

describing the response to an input impulse of area Φ_1. Comparing this result with equation (2.2.1) we finally write for the (normalized) impulse response $h(t)$,

$$h(t) = u(t) (1/RC) \exp (-t/RC). \tag{2.2.7}$$

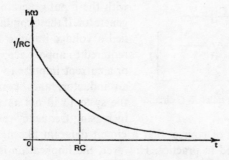

Figure 2.3b. Normalized impulse response of the system shown in figure 2.3a.

This function is plotted in figure 2.3b. Since $u(t)$ and $\exp (-t/RC)$ are dimensionless, the units along the y axis are those of inverse time. These units may seem more meaningful if they are expressed in the form 'voltage output per unit voltage impulse at the input' which has the dimensions of $1/t$.

If we had been considering the current $i_C(t)$ rather than the voltage $v_C(t)$ as the output quantity, we would have obtained an impulse response having the form of the derivative of equation (2.2.7). Since (2.2.7) has a discontinuity or 'step' at $t = 0$, its derivative will have an impulse at $t = 0$. Physically this corresponds to part of the input being transferred directly to the output, and it is often convenient to consider the impulsive part of the response separately from the rest. We shall see later that a system with an impulse response of this type has a frequency response (section 3.1) which does not fall off to zero in the high-frequency limit.

One can visualize systems whose impulse response contains derivatives of impulses; for example in figure 2.3c the output current $i(t)$ has a term proportional to the derivative of the input voltage $v(t)$. If $v(t)$ is to be an impulse function, we must suppose that the capacitor is instantaneously charged to an infinite energy and immediately discharged again. This infinite-energy requirement is sufficient to establish the fact that figure 2.3c is not a completely satisfactory model of a physical system. In frequency-response terms, it requires the output amplitude (for a given input amplitude) to increase indefinitely with increasing frequency. In fact, if a circuit diagram is to be a general representation of a physical system, it must always show some resistance in series with the input generator,

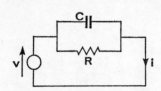

Figure 2.3c. Parallel *R–C* circuit.

if a voltage generator, or in parallel with the input generator, if a current generator. If these conditions are satisfied a voltage impulse will never be required to appear across a capacitor or a current impulse to flow through an inductor, and, correspondingly, the system will not have an infinitely increasing frequency-response.

It is true, of course, that a circuit diagram like figure 2.3c may often be used in practice. However, the implication is not that the voltage generator has absolutely zero output resistance; in fact, every physical generator has a finite output resistance – in other words, every physical generator dissipates energy when it delivers power into a load. In practical applications, we imply when we draw a circuit like figure 2.3c that the frequency components of the generator are restricted to a range in which the effect of its output resistance is negligible, and this restriction prevents us from considering the impulse response. Where the circuit diagram shows a resistance in series with the voltage generator, as in figure 2.3a, there is no such difficulty because this resistance can be assumed to include the output resistance of the source.

At this point it is interesting to refer back to the discussion of stability and causality in section 1.2, because the impulse response function $h(t)$ can be used to provide a simple formulation of these principles in the following way:

$$\text{Stability:} \int_{-\infty}^{\infty} |h(t)| \mathrm{d}t < \infty . \tag{2.2.8}$$

$$\text{Causality:} \ h(t) = 0 \text{ for } t < 0 . \tag{2.2.9}$$

2.3 The convolution integral

It has already been mentioned that if we know the impulse response $h(t)$ of a linear system, we can deduce the response to any input function whatever.

To demonstrate this fact, let us consider the input function $x(t)$ shown in figure 2.4.

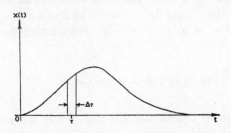

Figure 2.4. An input function $x(t)$.

We split up $x(t)$ into a succession of rectangular elementary pulses, each of width $\Delta\tau$, so that the area of a typical pulse at $t = \tau$ is $x(\tau)\Delta\tau$. As we have seen in section 2.2, in the limit as $\Delta\tau \to 0$ this pulse gives rise to a response $[x(\tau)\Delta\tau]h(t-\tau)$. The total response $y(t)$ may therefore be expressed as the integral

$$y(t) = \int\limits_{-\infty}^{\infty} x(\tau)h(t-\tau)\mathrm{d}\tau . \tag{2.3.1}$$

Notice the use of the two symbols t and τ for the time variable. τ is often called the *dummy variable* because it disappears after the integration has been performed. The physical meaning of the integral can be expressed by the statement that the value of the output at a given time t is the integrated effect of the values of the input at all previous times τ; this brings out the difference between the time variable t and the time variable τ. It is also worth noting that the upper limit of integration can be changed from ∞ to t because the impulse response $h(t-\tau)$ has zero value for $\tau > t$.

Equation (2.3.1) defines the *convolution integral* of $x(t)$ and $h(t)$. Any two functions of the same variable can be combined in this particular way which may be regarded as a kind of multiplication and expressed by the special symbol $\otimes$. The following general statement defines $\otimes$, known as the *convolution operation*:

$$a(t) \otimes b(t) = \int\limits_{-\infty}^{\infty} a(\tau)b(t-\tau)\mathrm{d}\tau . \qquad (2.3.2)$$

An important property of the convolution operation may be demonstrated by means of a simple change of variable. Putting $(t-\tau) = \tau'$, we have $\tau = (t-\tau')$ and $\mathrm{d}\tau = -\mathrm{d}\tau'$. The limits $\tau = -\infty$ to $\tau = \infty$ now become $\tau' = \infty$ to $\tau' = -\infty$. Substituting these quantities we obtain

$$a(t) \otimes b(t) = -\int\limits_{\infty}^{-\infty} a(t-\tau')b(\tau')\mathrm{d}\tau' . \qquad (2.3.3)$$

Comparison of equation (2.3.2) and (2.3.3) brings out the fact that the convolution operation obeys the *commutation rule*

$$a(t) \otimes b(t) = b(t) \otimes a(t) \qquad (2.3.4)$$

and it may also be shown that the operation has the properties of *distribution* and *association*, defined by equations (2.3.5) and (2.3.6) respectively:

$$a(t) \otimes [b(t) + c(t)] = a(t) \otimes b(t) + a(t) \otimes c(t) \qquad (2.3.5)$$

$$a(t) \otimes [b(t) \otimes c(t)] = [a(t) \otimes b(t)] \otimes c(t) . \qquad (2.3.6)$$

Having established the meaning of the convolution symbol, let us refer back to equation (1.1.2) which introduced the system operator $\mathscr{S}$. Using the operator notation, equation (2.3.1) has the form

$$y(t) = x(t) \otimes h(t) = h(t) \otimes x(t) \qquad (2.3.7)$$

so that by a direct comparison with equation (1.1.2) we can see that one of the ways in which $\mathscr{S}$ can be given a specific form is

$$\mathscr{S} = h(t) \otimes . \qquad (2.3.8)$$

The fact that $\mathscr{S}$ transforms a real function into another real function is seen to be a mathematical consequence of the fact that $h(t)$ must be real.

2.4 Step response

The simplest application of the convolution integral is in the calculation of the response of a linear system to an input of the form $x_1u(t)$. By taking the ratio of this response to the magnitude of the step-function which caused it, we may define a *normalized step response* $a(t)$:

$$a(t) = (1/x_1)\mathscr{S}[x_1u(t)] . \qquad (2.4.1)$$

$a(t)$ is usually referred to simply as the *step response*.

By means of the alternative statement for $\mathscr{S}$ given by equation (2.3.8) we may rewrite equation (2.4.1) in the form

$$a(t) = h(t)\otimes u(t) = \int_{-\infty}^{\infty} h(\tau)u(t-\tau)\mathrm{d}\tau . \qquad (2.4.2)$$

Now since the system is causal we may apply equation (2.2.9) which states that $h(\tau) = 0$ for $\tau < 0$. Also, we have, $u(t-\tau) = 0$ for $\tau > t$, so, provided that $t > 0$, the effective range of integration must be from $\tau = 0$ to $\tau = t$. Now the value of $u(t-\tau)$ is defined as unity within these limits so that the final expression for the step response has the following simple form:

$$a(t) = \int_{0}^{t} h(\tau)\mathrm{d}\tau . \qquad (2.4.3)$$

Thus it is seen that *the step response of a linear system is the integral of the impulse response*, a fact which is obviously related to the step function being the integral of the impulse function. At a later stage we shall examine a general principle for linear systems which may be summarized by the statement that if the input is differentiated or integrated, so is the response; the case described here is an example of this general principle.

An important special case of (2.4.3) is obtained by substituting $t = \infty$:

$$a(\infty) = \int_{0}^{\infty} h(t)\mathrm{d}t . \qquad (2.4.4)$$

Equation (2.4.4) shows that the time integral of the impulse response of a linear system is equal to the 'static' value of the step response, that is to the steady value approached by the step response as $t \to \infty$. Thus for the system shown in figure 2.3a the static value of $a(t)$ is obviously unity, which is also the value of the time integral of the impulse response given in equation (2.2.7).

Equation (2.4.3) is often used for calculating the impulse response from the step response, which is often easier to measure in practice.

2.5 Complex exponential input

Apart from the impulse function $\delta(t)$ and the closely related step function $u(t)$ there is another basic input function which is of primary interest in the study of linear systems. This is the *complex exponential* function:

$$x(t) = x_1 \exp st \qquad (2.5.1)$$

where x_1 and s are complex quantities; (2.5.1) is, in general, an oscillatory function with exponentially decaying or increasing amplitude, and we can split it up into real and imaginary parts by writing

$$x_1 = |x_1| \exp j\alpha, \quad s = \sigma + j\omega \qquad (2.5.2)$$

so that,

$$x(t) = |x_1| \exp \sigma t \left[\cos (\omega t + \alpha) + j \sin (\omega t + \alpha) \right]. \qquad (2.5.3)$$

The parameter s is sometimes called the 'complex frequency', but we must always bear in mind that, in accordance with equation (2.5.2), the familiar angular frequency ω is the *imaginary* part of the complex quantity s.

If ω has a non-zero value, the single term in (2.5.1) cannot be the actual input function of any physical system; any complex term must always be accompanied by another term which is its conjugate, because of the requirement that $x(t)$ for a physical system must be real. However, when analyzing the mathematical properties of the system operator we are subject to no such restrictions, and we shall now consider the effect of a linear operator $\mathscr{S}$ on the input function $x_1 \exp st$ in a purely formal way. By using the principle of convolution, we may express the output $y(t)$ in terms of the impulse response $h(t)$. We have,

$$y(t) = \mathscr{S}[x_1 \exp st] = x_1 h(t) \otimes \exp st$$

$$= x_1 \int_{-\infty}^{\infty} h(\tau) \exp s(t-\tau) \, d\tau. \qquad (2.5.4)$$

Assuming that this integral converges, we can extract the time-dependent factor and write

$$y(t) = H(s) x_1 \exp st \qquad (2.5.5)$$

where

$$H(s) = \int_{-\infty}^{\infty} h(t) \exp -st \, dt . \qquad (2.5.6)$$

We have been able to dispense with the symbol τ, as only one time variable is present in the integral.

We must now consider the convergence of the integral in (2.5.6). The integral

$$\int_{-\infty}^{\infty} h(t) \exp -j\omega t \, dt \qquad (2.5.7)$$

is the Fourier transform of $h(t)$, and will converge for all values of ω if $h(t)$ is the impulse response of a stable system as defined by equation (2.2.8). We now write the integral in equation (2.5.6) in the form

$$\int_{-\infty}^{\infty} h(t) \exp -j\omega t \exp -\sigma t \, dt . \qquad (2.5.8)$$

If σ is positive, then $\exp -\sigma t$ is always less than unity for positive values of t. Now if the system is causal, we know from equation (2.2.9) that $h(t)$ is zero for negative values of t. In this case (2.5.8) is less than (2.5.7) and therefore must also converge if (2.5.7) converges.

The conclusion is that the *integral in equation* (2.5.6) *has finite values for $\sigma \geqslant 0$, provided that the system is stable and causal.*

If the integral converges, equation (2.5.5) shows that the output will then have the same time-dependence as the input, say $y_1 \exp st$, and the relation between the input and the output can therefore be represented by the ratio $y_1/x_1 = H(s)$ which is a complex quantity independent of time. $H(s)$ is called the *system function*, and we shall see shortly that it provides a method of characterizing a linear system which is often much more convenient than the impulse response $h(t)$.

2.6 The system differential equation

Suppose that the physical relationship between the input quantity x and the output quantity y can be expressed in the form of the differential equation

$$A_0 y + A_1 (dy/dt) + A_2 (d^2y/dt^2) + \dots$$
$$= B_0 x + B_1 (dx/dt) + B_2 (d^2x/dt^2) + \dots . \qquad (2.6.1)$$

Now the solution $y(t)$ of this equation can be expressed as the sum of two functions

$$y(t) = y_c(t) + y_p(t) \qquad (2.6.2)$$

where $y_p(t)$ is any function which satisfies (2.6.1). The function $y_c(t)$ is called the *complementary function* in textbooks of mathematics, but we shall call it the *free response*; it is a solution of the equation

$$A_0 y + A_1 (dy/dt) + A_2 (d^2y/dt^2) + \ldots = 0 \qquad (2.6.3)$$

and it is chosen in such a way that (2.6.2) satisfies the required boundary conditions. To solve equation (2.6.3) we set up the auxiliary equation

$$A_0 + A_1 p + A_2 p^2 + \ldots = 0. \qquad (2.6.4)$$

The roots of this equation are $p_1, p_2 \ldots p_n$ where n is the order of equation (2.6.3). If all these roots are different, the required solution is

$$y_c(t) = C_1 \exp p_1 t + C_2 \exp p_2 t + C_3 \exp p_3 t + \ldots \qquad (2.6.5)$$

where the coefficients are determined by the boundary conditions, for instance by specifying the values of y and some of its derivatives for $t = 0$.

If two of the roots p_1 and p_2 are equal, the solution takes the form

$$y_c(t) = (C_1 + C_2 t) \exp p_1 t + C_3 \exp p_3 t + \ldots \qquad (2.6.6)$$

Similarly a root repeated three times gives an additional term in $t^2 \exp p_1 t$ and so on. Where roots are complex, they occur in conjugate pairs, each giving rise to a pair of complex-exponential terms in $y_c(t)$ whose coefficients are also conjugate. These two terms combine to give a single term of the form $\exp \sigma t \cos (\beta t + \alpha)$ where σ is the real part, and $\pm \beta$ the imaginary part, of each root.

The second part $y_p(t)$ of the response is called the *particular integral*, and its form depends on the form of the input function $x(t)$.

When we are considering the response of the system to a switched sinusoidal input $x_1 u(t) \cos \omega_1 t$, we can easily assign a physical meaning to the two parts of the solution. For $t > 0$ the input function is $x_1 \cos \omega_1 t$ and the particular integral is the response which we would calculate for an input of this kind; direct substitution in the equation shows that it is a sinusoidal function of angular frequency ω_1. To this we must add free-response terms which will enable us to satisfy the required boundary conditions at $t = 0$; these terms constitute the *switching transient*. After the switching transient has died away, only the particular integral remains, and for this reason it is often called the *forced response* in this context.

In section 2.5 we have defined $H(s)$ by considering an input function $x_1 \exp st$ which gives rise to an output $y_1 \exp st$ provided that the integral (2.5.6) converges. $H(s)$ is the complex ratio y_1/x_1. Substituting these input and output functions in the original differential equation (2.6.1) and taking out the factor $\exp st$, we obtain

$$y_1(A_0 + A_1 s + A_2 s^2 + ..) = x_1(B_0 + B_1 s + B_2 s^2 + ..) \quad (2.6.7)$$

$$H(s) = y_1/x_1 = \frac{B_0 + B_1 s + B_2 s^2 + \cdots}{A_0 + A_1 s + A_2 s^2 + \cdots}. \quad (2.6.8)$$

This result leads to the very important conclusion that, for the large class of systems which are described by a linear differential equation with constant coefficients of the form given in equation (2.6.1), $H(s)$ is simply a rational function of s with coefficients directly obtained from the differential equations.

An electrical circuit containing a finite number of linear components can be described by an equation of the form of (2.6.1), and $H(s)$ for such a circuit is therefore a rational function of s. However, $H(s)$ for such a system can be immediately written down, without the need for explicitly setting up the differential equation. This is achieved by the familiar method of *complex impedances* where each inductor L is represented by an impedance sL, each capacitor C by an impedance $1/sC$, and each resistor R by an impedance R. The relation between the output quantity (voltage or current) and the input quantity (voltage or current) is then expressed by the use of circuit theorems analogous to those used in d.c. circuits. Thus for the circuit shown in figure 2.5, where the input is the voltage v_1 and the output the voltage v_2, we may immediately apply the 'potential divider' principle and write

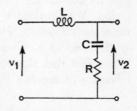

Figure 2.5. Series $L-R-C$ circuit.

$$H(s) = \frac{R + 1/sC}{R + sL + 1/sC} = \frac{s + 1/RC}{s^2 L/R + s + 1/RC}. \quad (2.6.9)$$

By substituting $j\omega$ for s we obtain $H(j\omega)$, the frequency-response function; we note that in circuits where the output falls to zero as the frequency approaches infinity the order of the denominator of $H(s)$ is higher than the order of the numerator. This applies to most of the systems in which we are interested.

In cases where $H(s)$ is a rational function of s, it is convenient to factorize it in the form

$$H(s) = \text{constant} \times \frac{(s-z_1)(s-z_2)\ldots}{(s-p_1)(s-p_2)\ldots}. \qquad (2.6.10)$$

The complex quantities p_1, p_2 etc. are called the *poles* of the function $H(s)$, and are in fact the same parameters that were used in equations (2.6.5) and (2.6.6) when we were discussing the free response. Since they are the roots of the equation obtained by putting the denominator of $H(s)$ equal to zero, all complex poles occur in conjugate pairs. The quantities $z_1, z_2 \ldots$ are called the *zeros* of $H(s)$ and, like the poles, are real or in complex conjugate pairs. If any factor in the numerator or denominator is repeated n times, the corresponding zero or pole is said to be of order n.

Apart from a constant factor, the function $H(s)$ is completely specified by its poles and zeros, so that a convenient way of 'plotting' $H(s)$ is to indicate the position of these points in the complex plane. Now we have seen in section 2.5 that the defining integral (2.5.6) is finite for $\sigma > 0$ provided that the system is stable and causal. This implies that $H(s)$ is finite for all values of s on the $j\omega$ axis and in the right-hand half of the complex s-plane, so that its denominator does not become zero at any point in this region. We thus reach the very important conclusion that *for a stable causal system, $H(s)$ can have no poles in the right-hand half of the complex s-plane.*

The fact that the poles and zeros are in conjugate pairs is a reflection of the fact that $H(s)$ has conjugate symmetry, i.e. that

$$H(s^*) = H^*(s) \qquad (2.6.11)$$

This result can easily be obtained from the defining integral (2.5.6) with the assumption that $h(t)$ is real. It implies that $H(s)$ is real at all points on the σ axis.

Notice that whenever $H(s)$ is plotted, or whenever numerical values are substituted, it is necessary to indicate the units which have been used. The physics-trained reader is likely to be confused by the formulation used by authors of engineering texts, who do not in general make any attempt to ensure that their formulae are dimensionally correct. It is usually best to rewrite the formulae in a dimensionally correct form; for instance the function $(s+1)/(s+2)(s+3)$ means $(s+\omega_1)/(s+2\omega_1)(s+3\omega_1)$ where ω_1 is of course the appropriate normalizing frequency.

B

CHAPTER 3

USE OF FOURIER AND LAPLACE TRANSFORMS

3.1 Fourier series and Fourier transform

In section 2.5 we introduced the system-function $H(s)$ by considering the response of a system to an input of the type

$$x(t) = x_1 \exp st \qquad (3.1.1)$$

where $s = \sigma + j\omega$.

It is most important to notice that we have defined ω as a real quantity, being the imaginary part of the complex quantity s. This is the usual definition in texts on electronics and control, and will be used throughout this book. On the other hand, in texts on physics ω is sometimes treated as a complex quantity.

We now have to discuss the function $H(s)$ in much more detail, and in this section we shall restrict ourselves by assuming that σ is identically equal to zero, and that the system is stable. Instead of $H(s)$ we now write $H(j\omega)$, and this function may be called the *frequency response* of the system. $H(j\omega)$ is of course a function of the real variable ω, and our only reason for writing $j\omega$ in the bracket rather than ω is to enable us to use the same symbol H as before. From (2.5.6) we have the definition

$$H(j\omega) = \int_{-\infty}^{\infty} h(t) \exp -j\omega t \, dt . \qquad (3.1.2)$$

This is a statement of the very important fact that *the frequency response is the Fourier transform of the impulse response.* The transform will exist (i.e. the integral in (3.1.2) will converge) if the system is stable. Now this definition does not restrict us to positive values of ω, and because $h(t)$ is real we see that

$$H(j\omega) = H^*(-j\omega) \qquad (3.1.3)$$

24

which is a special case of the conjugate-symmetry property of $H(s)$ given in equation (2.6.11).

It follows from the discussion in section 2.5 that if the input function is assumed to be of the form

$$x(t) = x_1 \exp j\omega t \qquad (3.1.4)$$

then the calculated output will be

$$y(t) = H(j\omega)x_1 \exp j\omega t . \qquad (3.1.5)$$

In order to establish the physical meaning of $H(j\omega)$, we consider a real input function of the form

$$x(t) = 2x_1 \cos \omega_1 t \qquad (3.1.6)$$

where x_1 is real and ω_1 is implicitly real and positive. This can be expressed as the sum of two terms of the type defined in equation (3.1.4); in one of these terms the value of ω is ω_1, and in the other it is $-\omega_1$. Notice that when we are defining a physical input it is not meaningful to distinguish between positive and negative frequency; on the other hand, there is a clear distinction in the case of these complex terms because the sign of ω determines the direction of rotation in the complex plane. Now $H(j\omega)$ has been defined to include both positive and negative ranges of ω, and accordingly we use $H(j\omega_1)$ and $H(-j\omega_1)$ to calculate the operation of the system on the two complex terms in the input:

$$y(t) = H(j\omega_1)x_1 \exp j\omega_1 t + H(-j\omega_1)x_1 \exp -j\omega_1 t . \qquad (3.1.7)$$

Referring to (3.1.3) we see that the two terms in (3.1.7) are conjugates, a necessary requirement if $y(t)$ is to be real. We can therefore rewrite (3.1.7) in the form

$$y(t) = Re [H(j\omega_1) 2x_1 \exp j\omega_1 t] \qquad (3.1.8)$$

where the operator Re indicates that only the real part is taken. Now if we put $H(j\omega_1)$ in the polar form

$$H(j\omega_1) = A(\omega_1) \exp j\theta(\omega_1) \qquad (3.1.9)$$

we can express (3.1.8) as follows:

$$y(t) = 2x_1 A(\omega_1) \cos [\omega_1 t + \theta(\omega_1)] . \qquad (3.1.10)$$

The output is a sinusoidal function of the same frequency as the input, the amplitude and phase relations between output and input being given by the magnitude and angle of the frequency-response function.

In principle it is impossible to have an input to a physical system which is exactly of the form of equation (3.1.6), because this describes a function which goes on for all time. In practice if we wish to measure $H(j\omega)$ at a given frequency we simply switch on an input from a sinusoidal generator, and wait until the switching operation has been 'forgotten' and the output has settled down to a sinusoidal function of constant amplitude. In general, whenever we say that the time variation of a physical quantity is periodic we do not mean that it has been going on forever – simply that it has been going on long enough to be indistinguishable in practice from an ideally periodic variation.

Now the usefulness of $H(j\omega)$ is not restricted to cases where the input is a sinusoidal function of time. For non-sinusoidal inputs we can make use of the frequency-response function by means of the Fourier series or the Fourier transform, depending upon the nature of the input.

(i) A periodic input $x(t) = x(t+\tau_1)$ can be expressed in the form of a complex Fourier series,

$$x(t) = \sum_{n=-\infty}^{\infty} X_n \exp jn\omega_F t \qquad (3.1.11)$$

where the Fourier coefficients X_n and the fundamental frequency ω_F, are defined by the relations,

$$X_n = (1/\tau_1) \int_{-\tau_1/2}^{\tau_1/2} x(t) \exp -jn\omega_F t \, dt \qquad (3.1.12)$$

$$\omega_F = 2\pi/\tau_1 .$$

Equation (3.1.11) describes how the input $x(t)$ can be represented by a series of sinusoidal functions. Because the system is linear, each component $X_n \exp jn\omega_F t$ of the input gives rise to an output $H(jn\omega_F) X_n \exp jn\omega_F t$. We may therefore express the resulting output in the form

$$y(t) = \mathscr{S}x(t) = \sum_{n=-\infty}^{\infty} H(jn\omega_F) X_n \exp jn\omega_F t \qquad (3.1.13)$$

This expression gives $y(t)$ real for real $x(t)$, because in this case $X_n = X_{-n}^*$, and since also $H(j\omega) = H^*(-j\omega)$ [equation (3.1.3)] the summation is over pairs of complex conjugates.

Equation (3.1.13) can equally well be expressed in terms of the Fourier coefficients Y_n of the output $y(t)$:

$$Y_n = H(jn\omega_F) X_n \qquad (3.1.14)$$

and the result is equivalent to the following statement. *Each Fourier coefficient of the output is obtained by multiplying the corresponding Fourier coefficient of the input by the appropriate value of $H(j\omega)$.*
(ii) A non-periodic input $x(t)$ can be expressed in the form

$$x(t) = (1/2\pi) \int\limits_{-\infty}^{\infty} X(j\omega) \exp j\omega t \, d\omega \qquad (3.1.15)$$

where $X(j\omega)$ is the Fourier transform of $x(t)$, defined by the relation

$$X(j\omega) = \int\limits_{-\infty}^{\infty} x(t) \exp -j\omega t \, dt \,. \qquad (3.1.16)$$

This integral is finite for all values of ω provided that $x(t)$ satisfies the condition

$$\int\limits_{-\infty}^{\infty} |x(t)| dt < \infty \,.$$

Physically, (3.1.16) is equivalent to expressing $x(t)$ as a continuous spectrum of sinusoidal components, and we see that (3.1.15) and (3.1.16) are analogous to (3.1.11) and (3.1.12) respectively. For any given frequency ω_1, the element $X(j\omega_1)d\omega$ in the spectrum of $x(t)$ gives rise to a contribution $H(j\omega_1) X(j\omega_1)d\omega$ at ω_1 in the output spectrum, and the analogous result to (3.1.14) is

$$Y(j\omega) = H(j\omega) X(j\omega) \qquad (3.1.17)$$

where $Y(j\omega)$ is the Fourier transform of the output function $y(t)$.

This result is equivalent to the statement that *the Fourier transform of the output is obtained by multiplying the Fourier transform of the input by $H(j\omega)$.*

To obtain $y(t)$ explicitly, we apply the inverse Fourier transform and obtain the equation

$$y(t) = (1/2\pi) \int\limits_{-\infty}^{\infty} H(j\omega)X(j\omega) \exp j\omega t \, d\omega \qquad (3.1.18)$$

which is the corresponding equation to (3.1.13).

We have thus found a means of expressing the relation between the output and the input of a linear system without the use of convolution integrals. This will be discussed in more detail in following sections.

3.2 Laplace transform

We are now going to extend the scope of the transform method outlined in the last section by making use of the complex variable $s = \sigma + j\omega$. In setting up the transform of a function $f(t)$ we make use of the integral

$$\int_{-\infty}^{\infty} f(t) \exp -st \, dt . \tag{3.2.1}$$

For a given value of s, this integral may or may not converge; whether it does so is determined by the value of σ, and it can be shown that in general, the integral (3.2.1) will exist for values of s lying in the strip of the complex s–plane defined by

$$\gamma_1 < \sigma < \gamma_2 \tag{3.2.2}$$

where γ_1 and γ_2 are parameters which depend on the function $f(t)$.

Now in most applications of the integral (3.2.1) we shall be dealing with functions $f(t)$ which are zero for $t < 0$, and in this case the integration is essentially over positive values of t. There is then no upper limit to the values of σ for which the integral converges, so that it converges over the half plane defined by

$$\sigma > \gamma_1 . \tag{3.2.3}$$

As an example we may take the cases

$$\text{(i)} \quad f(t) = u(t) \exp \alpha t \ (\alpha \text{ real}) \tag{3.2.4}$$

for which integral (3.2.1) converges to $1/(s-\alpha)$ for $\sigma > \alpha$, and

$$\text{(ii)} \quad f(t) = u(t) \cos \omega t \tag{3.2.5}$$

where the integral is $s/(s^2 + \omega^2)$ for $\sigma > 0$.

Bearing in mind the convergence limits, we may define a function $F(s)$ such that

$$F(s) = \int_{-\infty}^{\infty} f(t) \exp -st \, dt \tag{3.2.6}$$

for $\gamma_1 < \sigma < \gamma_2 .$

Now although (3.2.6) defines $F(s)$ only in a limited region of the complex s–plane, we shall be treating $F(s)$ as a function of s which may exist in any part of the s–plane. For instance, $F(s)$ is in many cases a rational function of s, and although this will be equal to the defining integral only over a limited range of s, it obviously has a value for any s that we may choose. We shall refer to $F(s)$ as the (bilateral) *Laplace transform* of $f(t)$.

It could equally well be called the complex-frequency Fourier transform provided that one made it clear that the physical angular frequency ω is the imaginary part, rather than the real part, of s. However we shall reserve the name 'Fourier transform' for the function $F(j\omega)$ defined as in equation (3.1.16). Many authors make use of the *unilateral Laplace transform* in which the integral is taken from 0 to ∞ rather than from $-\infty$ to ∞; in this book we shall not use the unilateral transform (see preface).

Referring back to section 2.5, we see that the system function $H(s)$ is the Laplace transform of the impulse response $h(t)$. If the system is stable and causal, the defining integral converges everywhere in the right-half s–plane.

It is often convenient to use the operator form

$$F(s) = \mathscr{L} f(t) \tag{3.2.7}$$

and
$$f(t) = \mathscr{L}^{-1} F(s)$$

where the operator $\mathscr{L}^{-1}$ refers to the *inverse Laplace transform*. Formally, the inverse Laplace transform may be defined by the equation

$$f(t) = (1/2\pi j) \int_{\alpha - j\infty}^{\alpha + j\infty} F(s) \exp st \, ds. \tag{3.2.8}$$

Here, the path of integration is assumed to be the straight line parallel to the $j\omega$ axis defined by $\sigma = \alpha$, where α is a parameter whose value may be arbitrarily chosen provided that this line lies in the region of the s–plane in which the defining integral (3.2.6) converges. Equation (3.2.8) can be obtained immediately by considering the Fourier transform of the function $f(t) \exp -\alpha t$; in fact, in very many cases it is possible to put $\alpha = 0$, and then the inverse Laplace transform operation defined in (3.2.8) is identical with the inverse Fourier transform. When the integral is performed numerically, therefore, there is practically no difference between the two inverse transforms; on the other hand, we shall see that the analytical evaluation of $f(t)$ may be made very much easier by treating $F(s)$ as a function of a complex variable.

The fact that the Laplace operator $\mathscr{L}$ is a linear one is obvious from its definition. Two further properties which must be mentioned here are

$$\mathscr{L} \, \mathrm{d}f(t)/\mathrm{d}t = s \, \mathscr{L} f(t) \qquad (3.2.9)$$

and

$$\mathscr{L} \, [f_1(t) \otimes f_2(t)] = \mathscr{L} f_1(t) \, \mathscr{L} f_2(t) . \qquad (3.2.10)$$

Equation (3.2.9) is not meant to state that the transform of $\mathrm{d}f(t)/\mathrm{d}t$ necessarily exists, but that if it does exist it is given by $s \, \mathscr{L} f(t)$; the convergence limits of the two defining integrals are not generally the same. Results (3.2.9) and (3.2.10) can be derived from the defining integral (3.2.6), the first by partial integration, the second in conjunction with equation (2.3.2) which defines the convolution operation.

Table 3.1

$f(t)$	$F(s)$	Convergence
$\delta(t)$	1	everywhere
$\delta(t-\tau)$	$\exp -s\tau$	everywhere
$u(t)$	$1/s$	$\sigma > 0$
$u(t-\tau)$	$(1/s) \exp -s\tau$	$\sigma > 0$
$u(t+\tau)-u(t-\tau)$	$(1/s)(\exp s\tau -\exp -s\tau)$	everywhere (reduces to $(2 \sin \omega\tau)/\omega$ for $\sigma = 0$)
$u(t) \exp \alpha t$	$1/(s-\alpha)$	$\sigma > \alpha$
$u(t) \cos \alpha t$	$s/(s^2+\alpha^2)$	$\sigma > 0$
$u(t) \sin \alpha t$	$\alpha/(s^2+\alpha^2)$	$\sigma > 0$
$u(t)t^n$ (n any $+ve$ integer)	$n!/s^{n+1}$	$\sigma > 0$
$u(t)t^n \exp \alpha t$ (n any $+ve$ integer)	$n!/(s-\alpha)^{n+1}$	$\sigma > \alpha$

Table 3.1 gives a number of useful functions of t with their corresponding Laplace transforms and the regions of convergence of the defining integral. In the cases where the region of convergence includes the $j\omega$ axis, i.e. where the defining integral converges for $\sigma = 0$, the function has a Fourier transform obtained by substituting

$s = j\omega$ in the Laplace transform. Although the meaning of the transform is usually easier to visualize from a physical point of view when it is put in Fourier form, the Laplace formulation is needed when we wish to make use of the analytic properties of the transform.

We now proceed to establish the relation between the output and the input of a linear system in terms of the Laplace transform. Taking the convolution formula

$$y(t) = h(t) \otimes x(t) \qquad \qquad [(2.3.7)]$$

in conjunction with equation (3.2.10) we immediately reach the result

$$Y(s) = H(s) \, X(s) \qquad \qquad (3.2.11)$$

where $Y(s)$, $H(s)$, and $X(s)$ are the Laplace transforms of $y(t)$, $h(t)$, and $x(t)$ respectively.

Now in section 3.1, by considering the physical meaning of the Fourier transform, we have already arrived at the relation

$$Y(j\omega) = H(j\omega) \, X(j\omega) \qquad \qquad [(3.1.17)]$$

and equation (3.2.11) can, in most applications, be regarded as an alternative statement of this same result, with the special advantage of emphasizing the fact that the system function can be treated as a function of the complex variable s.

As a means of formulating the relation between $y(t)$ and $x(t)$, equation (3.2.11) [or (3.1.17)] has the advantage over (2.3.7) that it does not involve convolution integrals. Another advantage is that in practical cases we are likely to know $H(s)$ directly, either by consideration of the differential equations relating to the system, from the component values if it is an electrical circuit, or experimentally by measurement of the frequency response. However, the use of the transform formulation involves the necessity of obtaining the inverse transform of $Y(s)$ if $y(t)$ is required, and we shall see in the next section how we can perform this operation by making use of the properties of a complex variable.

3.3 The inversion integral
Before proceeding to discuss the inversion integral (3.2.8), we shall remind ourselves of some of the principles of contour integration.

A function $F(s)$ is said to be *analytic* at a point in the s–plane if it has a unique derivative $dF(s)/ds$ at that point. According to

Cauchy's Theorem, if $F(s)$ is analytic everywhere on and within a closed curve **C**, then

$$\int_C F(s)\mathrm{d}s = 0 . \qquad (3.3.1)$$

We are interested in functions $F(s)$ which are analytic except at isolated points p_1, p_2 ... and begin by examining the behaviour of $F(s)$ in the vicinity of one of these points, that is when s approaches the value p_1.

The point p_1 is said to be *a first-order pole of residue* K_1 if

$$F(s) \to K_1/(s-p_1) \text{ as } s \to p_1 . \qquad (3.3.2)$$

For the first-order pole p_1, therefore, we may define the residue K_1 by the formula

$$\text{first-order pole: } K_1 = [(s-p_1)F(s)]_{s=p_1} . \qquad (3.3.3)$$

In the case of an *nth order pole* p_i (where $n > 1$), the expansion corresponding to equation (3.3.2) contains further terms in negative powers of $(s-p_i)$ up to $(s-p_i)^{-n}$. The residue K_i is still the co-efficient of the term in $(s-p_i)^{-1}$, but formula (3.3.3) is no longer applicable and must be replaced by the more general form

$$\text{nth order pole: } K_i = \frac{1}{(n-1)!}\left[\frac{\mathrm{d}^{n-1}}{\mathrm{d}s^{n-1}} (s-p_i)^n F(s) \right]_{s=p_i} . \qquad (3.3.4)$$

Now if we draw a closed path **C** which encloses poles p_1, p_2 ... of residues K_1, K_2, ... it can be shown that the line integral of $F(s)$ around **C** in the anticlockwise direction has the value

$$\int_C F(s)\mathrm{d}s = 2\pi\mathrm{j}(K_1+K_2+ \dots) . \qquad (3.3.5)$$

Equation (3.3.5) is called the *residue theorem* and is often rewritten

$$\int_C F(s)\mathrm{d}s = 2\pi\mathrm{j} \sum (\text{Residues of poles of } F(s) \qquad (3.3.6)$$
$$\text{enclosed by C}).$$

We shall apply this theorem to the evaluation of the inverse Laplace transform, starting with the derivation of the impulse response $h(t)$ from the system function $H(s)$. The required integral is

$$h(t) = \mathscr{L}^{-1} H(s) = (1/2\pi\mathrm{j}) \int_{\alpha-\mathrm{j}\infty}^{\alpha+\mathrm{j}\infty} H(s) \exp st \, \mathrm{d}s \qquad (3.3.7)$$

where the path of integration is assumed to be the straight line parallel to the jω axis defined by $\sigma = \alpha$, α being chosen so that this line lies in the region of convergence of the defining integral

$$H(s) = \int_{-\infty}^{\infty} h(t) \exp -st \, \mathrm{d}t. \qquad (3.3.8)$$

Another way of stating this criterion (assuming that the system is causal) is that α is greater than the real parts of any of the poles of $H(s)$.

We shall assume that we are dealing with a stable causal system so that all the poles in $H(s)$ lie in the left-hand half of the complex plane, and a suitable path of integration is $\alpha = 0$ (the jω axis). We also assume that, as $|s| \to \infty$, $H(s)$ approaches zero at least as fast as $1/s$ (this limiting condition is always true if $H(s)$ is a rational fraction of s with the denominator of higher order than the numerator); given that this condition is satisfied, the integral $\int H(s) \exp st \, \mathrm{d}s$ along a *semicircular* path from $s = \mathrm{j}R$ to $s = -\mathrm{j}R$ will approach zero as $R \to \infty$, *provided that the appropriate choice of semicircle is made according to the sign of t* (see figure 3.1). For $t < 0$, the path must lie in the *right-half-plane* and for $t > 0$ the path must lie in the *left-half-plane*, thereby ensuring that in both cases the product $[\sigma t]$ is always negative.

With this in mind we now consider the value of the integral $\int H(s) \exp st \, \mathrm{d}s$ around the closed contour $C_+ = $ AOBDA, shown

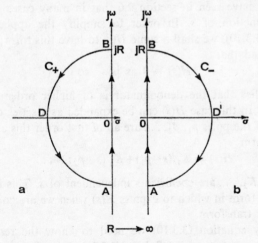

Figure 3.1. Contours for evaluation of $h(t)$.

in figure 3.1a, in the limit as $R \to \infty$. If $t > 0$, the contribution of the semicircular part BDA is zero and we may write

$$\int_{C_+} H(s) \exp st\, ds = \int_{-j\infty}^{j\infty} H(s) \exp st\, ds \text{ for } t > 0 \qquad (3.3.9)$$

where the integral on the right-hand side is the required result given by equation (3.3.7) for $\alpha = 0$.

Referring now to the general results expressed in equation (3.3.6) and (3.3.7) we see that

$$h(t) = \sum_{t>0} (\text{Residues of poles of } H(s) \exp st \text{ enclosed by } C_+). \qquad (3.3.10)$$

Applying the same principles to the contour $C_- = \text{AOBD'A}$ we arrive at the corresponding result for the right-half-plane, namely

$$h(t) = \sum_{t<0} (\text{Residues of poles of } H(s) \exp st \text{ enclosed by } C_-). \qquad (3.3.11)$$

The function $\exp st$ is analytic and non-zero for all values of s, so the poles of $H(s) \exp st$ are identical with those of $H(s)$; since $H(s)$ has no poles in the right-half-plane, equation (3.3.11) is equivalent to

$$h(t) = 0 \text{ for } t < 0 \qquad (3.3.12)$$

which is a statement of the causality property.

Now we have seen in section 2.6 that in many cases $H(s)$ is a rational function of s. In order to simplify the application of equation (3.3.10), we shall assume $H(s)$ to have this form; we have also assumed that

$$H(s) \to 0 \text{ as } |s| \to \infty$$

which implies that the denominator is of higher order than the numerator. In this case $H(s)$ can be expanded as a series of *partial fractions*; if the poles $p_1, p_2, \ldots$ are all of first order this expansion takes the form

$$H(s) = K_1/(s-p_1) + K_2/(s-p_2) + \ldots \qquad (3.3.13)$$

where $K_1, K_2, \ldots$ are coefficients independent of s. This is a very convenient form in which to express $H(s)$ when we are considering the inverse transform.

To apply equation (3.3.10) we need to know the residues of $H(s) \exp st$, and from the definition (3.3.3) we see that the residues at the poles p_1, p_2, ... are simply $K_1 \exp p_1 t$, $K_2 \exp p_2 t$, ... Thus from (3.3.10) and (3.3.12) we obtain the following result for the impulse response:

$$h(t) = u(t)\,(K_1 \exp p_1 t + K_2 \exp p_2 t + \dots)\,. \qquad (3.3.14)$$

We see that equation (3.3.14) has the same form as equation (2.6.5). This is because from a physical point of view the impulse response is simply the free response of the system with appropriate boundary conditions; the use of the transform gives us a simple way of evaluating the coefficients C_1, C_2, ...

If a pole p_1 lies on the real axis in the s–plane so that $p_1 = \sigma_1$, it gives rise to a simple exponential term $\exp \sigma_1 t$ in equation (3.3.14). If a pole p_2 is complex, such that $p_2 = (\sigma_2 + j\omega_2)$ it always has a conjugate partner $p_3 = (\sigma_2 - j\omega_2)$ and these poles together give rise to a term of the type $\exp \sigma_2 t \cos \omega_2 t$ in the impulse response.

In principle the system may have first-order poles for which $\sigma = 0$, and in this case the impulse response contains non-decaying sinusoids, or, for a pole actually at the origin, a constant term. Systems of this kind are not normally realizable in practice but they make useful theoretical models. They are sometimes classified as stable systems, but by our definition (2.2.8) they are unstable; a useful compromise is to classify them as *marginally stable*.

In calculations on actual systems it is usually convenient to make use of 'decay times' T_1, T_2, ... which are the reciprocals of the 'decay parameters' σ_1, σ_2, ... Physically, the decay time is the time taken for the exponential factor to fall to $1/e$ (37%) of its initial value. In numerical work it is important to remember that the units of T will be the reciprocal of the units which have been used for the values of the poles; thus if p_1, p_2 ... have been measured in kHz, the values of T_1, T_2 ... will be in milliseconds.

We must now consider the case where $H(s)$ has a pole of higher order than unity. The simplest example of this is

$$H(s) = K_a/(s - p_a)^2 \qquad (3.3.15)$$

where there is said to be a *second-order pole* at p_a. From (3.3.4) we see that the residue of $H(s) \exp st$ at the pole is $K_a t \exp p_a t$, and taking this result in conjunction with (3.3.10) and (3.3.12) we obtain for the impulse response

$$h(t) = u(t)\,K_a t \exp p_a t\,. \qquad (3.3.16)$$

This form of response has already been obtained in equation (2.6.6) by directly considering the solution of the system differential equation. In a practical case, the positions of the poles are likely to be determined ultimately by experimental measurements and the concept of a second-order pole really refers to the limiting case as two poles approach each other. The result given in (3.3.16) can in fact be obtained from (3.3.13) by considering this limiting case.

In general, if $H(s)$ has a pole of order m at p_a, the partial-fraction expansion will include fractions with denominators $(s-p_a)$, $(s-p_a)^2$, $\ldots (s-p_a)^m$. Because of the linearity of the inverse-transform operation, the impulse response $h(t)$ is simply the sum of the contributions from each partial fraction. Thus a second-order fraction gives rise to a term in $t \exp p_a t$ as in (3.3.16), a third-order fraction can similarly be shown to give a term in $t^2 \exp p_a t$, and so on.

So far we have concentrated on the calculation of the impulse response. The general problem in hand is to evaluate the inverse transform

$$y(t) = \mathscr{L}^{-1}\left[H(s)\,X(s)\right] \qquad (3.3.17)$$

where $X(s)$ is the transform of the input function $x(t)$; $x(t)$ is zero for $t<0$. Now we are not interested in input functions which increase indefinitely with time, and accordingly we shall assume that the input function is expressible in the form

$$x(t) = u(t)x_\mathrm{D}(t)+u(t)x_\mathrm{R}(t). \qquad (3.3.18)$$

Here $x_\mathrm{R}(t)$ is a periodic function with fundamental frequency ω_F, and $x_\mathrm{D}(t)$ is a function which approaches zero as $t \to \infty$; we shall assume that this approach is sufficiently rapid to ensure that the condition

$$\int\limits_0^\infty |x_\mathrm{D}(t)|\mathrm{d}t < \infty$$

is satisfied. If this is so, the region of convergence of the integral

$$\int\limits_{-\infty}^\infty u(t)x_\mathrm{D}(t)\exp -st\;\mathrm{d}t$$

will include the $j\omega$ axis; this implies that the poles of the transform of the first term in (3.3.18) are all in the left-half s–plane. The repetitive function $x_\mathrm{R}(t)$ can be expressed as a complex Fourier series having terms of the form $X_n \exp jn\omega_F t$, where the integer n takes both positive and negative values. The function $u(t)x_R(t)$ can therefore be expressed as a series of terms like $u(t)X_n \exp jn\omega_F t$, and from table 3.1 we see that each of these terms contributes a term $X_n/(s-jn\omega_F)$ to the transform. The poles of the transform of the second term in (3.3.18) are therefore on the $j\omega$ axis, and $x_R(t)$ being real they are in conjugate pairs.

It follows from this discussion that if the input function has the form indicated in equation (3.3.18) (obviously, in any particular case either $x_D(t)$ or $x_R(t)$ may be identically zero), the input transform $X(s)$ has no poles in the right-half s–plane. If the system is stable and causal, the product $H(s)X(s)$ also has no poles in the right-half s–plane. Now physically speaking the function $u(t)x_R(t)$ is indistinguishable from the decaying function $u(t)x_R(t) \exp \alpha t$, where α is real and negative, provided that α is sufficiently small in relation to the time of observation. We may use this argument to displace any imaginary poles of $H(s)X(s)$ by an infinitesimal distance into the left-half s–plane, so that they are now enclosed by the contour C_+ in figure 3.1. We now carry out the inverse transform in exactly the same way as we did in evaluating the impulse response, and accordingly find that for $t>0$ the output function $y(t)$ is equal to the sum of the residues of $H(s)X(s) \exp st$. It therefore contains exponential factors corresponding to the poles both of $H(s)$ and of $X(s)$, and this represents the fact stated in equation (2.6.2) that the total response can be expressed as the sum of a free response and a forced response. After a sufficient time has elapsed for the free response to decay to a negligibly small value, the response consists only of the residues of $H(s)X(s)$ at its imaginary poles; the same result would have been obtained by treating the input function as a purely repetitive one, that is as the function $x_R(t)$, and evaluating the response directly from equation (3.1.9) without using a transform at all.

If we know the response $y_1(t)$ to an input $x_1(t)$, we can easily calculate the response to the input $(d/dt)x_1(t)$. By equation (3.2.9), the transform of the new input function is $sX_1(s)$, and the required inverse transform is $\mathscr{L}^{-1}[sH(s)X_1(s)]$. Again by equation (3.2.9), we see that the response is $(d/dt)y_1(t)$. This result may be summarized by the statement that 'if the input is differentiated or integrated, so is the response'.

In the calculation of response functions it is sometimes convenient to make use of the step-response $a(t)$ which can be obtained from the expression

$$a(t) = \mathscr{L}^{-1}[H(s)/s] \qquad (3.3.19)$$

because $1/s$ is the transform of $u(t)$, or, equivalently, by direct use of the relation

$$a(t) = \int_0^t h(\tau)\mathrm{d}\tau . \qquad [(2.4.3)]$$

If an input can be regarded as a succession or combination of rectangular pulses we can make use of the relation

$$\mathscr{L}[u(t-\tau)] = \exp(-s\tau)/s \qquad (3.3.20)$$

to express the transform of the input as a series of exponential functions. The resulting expression for the response has terms of the form

$$\mathscr{L}^{-1}[H(s)\exp(-s\tau)/s] = (1/2\pi\mathrm{j}) \int_{-\mathrm{j}\infty}^{\mathrm{j}\infty} H(s)\exp s(t-\tau)\,\mathrm{d}s \qquad (3.3.21)$$

$$= a(t-\tau)$$

As we would expect, each step term at the input gives rise to its own step response, and the output is the sum of step responses with appropriate amplitudes and delays. It is usually much more convenient to make a single evaluation of $a(t)$ and employ the principle of superposition to find the output, rather than to go through the somewhat cumbersome procedure of making an explicit transform of the input function.

In the case of other forms of input function, it may be possible to make use of tables to find the required transform and inverse transform. Another approach is to calculate first the impulse response $h(t)$ and then make a numerical evaluation of the convolution integral (2.3.1).

3.4 Second-order system
We conclude this chapter by considering a very important case, that of the system described by a second-order differential equation. An example of such a system is the series $L–R–C$ circuit shown in figure 3.2, the input being a series voltage generator and the output the voltage across the capacitor C.

By using the complex-impedance method we can immediately write down the system function $H(s)$

$$\left.\begin{aligned} H(s) &= (1/sC)/(sL+R+1/sC) \\ &= \omega_N^2/(s^2+2\zeta\omega_N s+\omega_N^2) \end{aligned}\right\} \qquad (3.4.1)$$
$$\text{where } \omega_N = 1/\sqrt{(LC)}, \; \zeta = R/2\omega_N L \,.$$

The quantity ω_N may be called the *undamped natural frequency* of the system, and ζ the *damping ratio*. The poles of $H(s)$ are obtained by factorizing the denominator and are located at

$$-\zeta\omega_N \pm \omega_N \sqrt{(\zeta^2-1)} \,. \qquad (3.4.2)$$

It is instructive to consider the way in which the poles move as L and C are kept constant and R is increased from zero, that is as ω_N is kept constant and ζ is increased from zero.

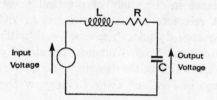

Figure 3.2. Series L–R–C circuit.

If ζ were zero, the poles would be imaginary and located at $\pm j\omega_N$, so that the impulse response of the system would be an undamped sinusoidal oscillation of frequency ω_N. As ζ increases towards unity, the poles move along the semicircular loci shown in figure 3.3 and the impulse response has the form of a damped

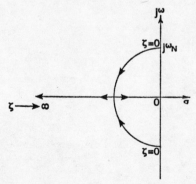

Figure 3.3. The loci of the poles of a series L–R–C circuit as ζ is varied.

sinusoidal oscillation with angular frequency $\omega_N \sqrt{(1-\zeta^2)}$ and decay parameter $\zeta\omega_N$. For the case where $\zeta^2 \ll 1$ the frequency of oscillation is very close to ω_N and the decay is comparatively slow. In this region it is often convenient to make use of the fact that the amplitude of oscillation falls to $1/e$ of its initial value in the course of $1/2\pi\zeta$ cycles of oscillation.

The condition $\zeta = 1$ is called the condition of *critical damping*. When $\zeta > 1$ the poles are both real, and, as ζ further increases, one pole approaches the origin ($s = 0$) and the other is at approximately $-2\zeta\omega_N$.

By substituting in equation (3.4.1) we can show that if $\zeta \geqslant 1/\sqrt{2}$ there is no peak in the frequency response, i.e. $|H(j\omega)|$ decreases monotonically as ω increases.

The characteristics of the idealized second-order system are of great importance in the study of practical linear systems, particularly with reference to such parameters as 'overshoot' in the time and frequency domains, 'rise time', and 'settling time' within a specified tolerance band. Computer-calculated curves of these quantities as a function of the damping ratio ζ have been published by P. Atkinson in *Feedback Control Theory for Engineers*, (Heinemann 1968).

CHAPTER 4

FREQUENCY-RESPONSE FUNCTIONS

4.1 Application of the Fourier transform

We have seen that the study of linear systems leads directly to the use of the Laplace transform in conjunction with the system function $H(s)$, where s is the complex-frequency variable. However, where possible it may be more convenient to work with the real variable ω, in which case we use the Fourier transform or Fourier series in conjunction with the frequency response function $H(j\omega)$. From a mathematical point of view the latter should be regarded as a special case of $H(s)$, and in fact we derive many of its analytic properties from a consideration of $H(s)$.

The Fourier transform $F(j\omega)$ of the time function $f(t)$ exists (i.e. the region of convergence of the Laplace transform integral includes the $j\omega$ axis) if $f(t)$ is absolutely integrable, that is if

$$\int_{-\infty}^{\infty} |f(t)|\,dt < \infty .$$

Obviously, the differentiation property (3.2.9) and the convolution property (3.2.10) of the Laplace transform apply equally to the Fourier transform. Because the Fourier transform integral and the corresponding inverse-transform integral are symmetrical (apart from a factor of 2π) between ω and t, these properties can be applied to the Fourier transform in the frequency domain as well as in the time domain. Thus if $F_1(j\omega)$ and $F_2(j\omega)$ are the transforms of $f_1(t)$ and $f_2(t)$ respectively, then the transform of $f_1(t)f_2(t)$ is $F_1(j\omega) \otimes F_2(j\omega)/2\pi$, and the inverse-transform of $dF_1(j\omega)/d\omega$ is $jtf_1(t)$.

41

Table 4.1. *Real and imaginary components of the Fourier transform of a real function $f(t)$.*

$$
\left. \begin{aligned}
F(j\omega) &= \int_{-\infty}^{\infty} f(t) \exp{-j\omega t}\, dt = F^*(-j\omega) \\
&= F_R(\omega) + jF_I(\omega)
\end{aligned} \right\}
\tag{4.1.1}
$$

$$
F_R(\omega) = \int_{-\infty}^{\infty} f(t) \cos \omega t\, dt = F_R(-\omega)
\tag{4.1.2}
$$

$$
F_I(\omega) = -\int_{-\infty}^{\infty} f(t) \sin \omega t\, dt = -F_I(-\omega)
\tag{4.1.3}
$$

$$
\text{if } f(t) = f(-t) \begin{cases} F_R(\omega) = 2\int_{0}^{\infty} f(t) \cos \omega t\, dt \\ F_I(\omega) = 0 \end{cases}
\tag{4.1.4}
$$

$$
\text{if } f(t) = -f(-t) \begin{cases} F_R(\omega) = 0 \\ F_I(\omega) = -2\int_{0}^{\infty} f(t) \sin \omega t\, dt \end{cases}
\tag{4.1.5}
$$

Inversion

$$
f(t) = (1/2\pi) \int_{-\infty}^{\infty} F(j\omega) \exp{j\omega t}\, d\omega
\tag{4.1.6}
$$

$$
= (1/2\pi) \int_{-\infty}^{\infty} (F_R(\omega) \cos \omega t - F_I(\omega) \sin \omega t)\, d\omega
\tag{4.1.7}
$$

$$
= (1/\pi) \int_{0}^{\infty} (F_R(\omega) \cos \omega t - F_I(\omega) \sin \omega t)\, d\omega
\tag{4.1.8}
$$

$$
= (1/\pi)\, Re \int_{0}^{\infty} F(j\omega) \exp{j\omega t}\, d\omega
\tag{4.1.9}
$$

In table 4.1 are given some useful relations involving the real and imaginary parts of the Fourier transform of a real function $f(t)$.

We shall now consider some properties of frequency-response functions.

4.2 Fourier transform of a causal function

It will be clear from the previous section that the only restrictions on $F_R(\omega)$ and $F_I(\omega)$, if they are to be the real and imaginary parts of the Fourier transform of a real function $f(t)$, are that $F_R(\omega)$ should be even and that $F_I(\omega)$ should be odd. We shall now show that if $f(t)$ is a causal function, defined by the property

$$f(t) = 0 \text{ for } t < 0$$

then $F_I(\omega)$ is uniquely defined by $F_R(\omega)$ and vice-versa.

Suppose that we are given $F_R(\omega)$ and wish to calculate $F_I(\omega)$. Because $f(t)$ is causal we can change the limits of the integral defining $F_R(\omega)$ and write

$$F_R(\omega) = \int_0^\infty f(t) \cos \omega t \, dt . \qquad (4.2.1)$$

Now let us define an even function $f_E(t)$ by the relations

$$\left. \begin{array}{l} f_E(t) = f(t) \quad , t > 0 \\ f_E(t) = f(-t), t < 0 \end{array} \right\} . \qquad (4.2.2)$$

The real and imaginary parts of the transform of this function are given, as in (4.1.4), by

$$\left. \begin{array}{l} F_{ER}(\omega) = 2 \int_0^\infty f_E(t) \cos \omega t \, dt = 2F_R(\omega) \\ F_{EI}(\omega) = 0 . \end{array} \right\} \qquad (4.2.3)$$

Now applying the inverse-transform formula, (4.1.8) we obtain an explicit expression for $f_E(t)$;

$$f_E(t) = (2/\pi) \int_0^\infty F_R(\omega) \cos \omega t \, d\omega . \qquad (4.2.4)$$

Because of the causality of $f(t)$ we may now write

$$F_I(\omega) = - \int_0^\infty f(t) \sin \omega t \, dt$$

$$= - \int_0^\infty f_E(t) \sin \omega t \, dt \tag{4.2.5}$$

and equations (4.2.4) and (4.2.5) show how $F_I(\omega)$ can be calculated if $F_R(\omega)$ is known. Similarly $F_R(\omega)$ can be obtained from $F_I(\omega)$ by means of the odd function $f_O(t)$:

$$\left. \begin{array}{ll} f_O(t) = f(t), & t > 0 \\ f_O(t) = -f(-t), & t < 0 \end{array} \right\} \tag{4.2.6}$$

$$\left. \begin{array}{l} f_O(t) = (-2/\pi) \displaystyle\int_0^\infty F_I(\omega) \sin \omega t \, d\omega \\[3mm] F_R(\omega) = \displaystyle\int_0^\infty f_O(t) \cos \omega t \, dt \end{array} \right\} . \tag{4.2.7}$$

Because the frequency-response function of a linear system is the Fourier transform of the impulse response, the results given above apply to the real and imaginary parts of the frequency-response function of a causal system.

The relationship is more conveniently expressed by the Kramers-Kronig relations which will be discussed in section 5.5.

4.3 All-pass, low-pass, and symmetrical band-pass systems

We shall now discuss the response of some linear systems in terms of the functions $A(\omega)$ and $\theta(\omega)$, defined by the relation

$$H(j\omega) = A(\omega) \exp j\theta(\omega) .$$

It will be convenient to consider the frequency range as covering both positive and negative values of ω, and we note that because the impulse response is real, $A(\omega)$ is an even function, and $\theta(\omega)$ an odd function, of ω. [See equation (3.1.3)].

Where $H(j\omega)$ is a rational function, $\tan \theta$ approaches zero or $\pm \infty$ in the high-frequency limit. The corresponding limiting value of θ, which is found by sketching out the function from $\omega = 0$ (where $\theta = 0$) to ∞, is always negative if $H(s)$ is a stable causal function which approaches zero in the high-frequency limit. A negative phase angle is called a *phase lag*.

We define a *distortionless* system as one in which the output $y(t)$ is an exact replica of the input apart from a *delay time* T_S and a scaling factor A_O, so that

$$y(t) = A_O x(t - T_S). \tag{4.3.1}$$

It follows from this that the impulse response and frequency-response function of such a system are

$$\left.\begin{aligned} h(t) &= A_O \delta(t - T_S) \\ H(j\omega) &= A_O \exp -j\omega T_S \end{aligned}\right\} . \tag{4.3.2}$$

The amplitude and phase angle functions are therefore given by

$$\left.\begin{aligned} A(\omega) &= A_O \\ \theta(\omega) &= -\omega T_S \end{aligned}\right\} . \tag{4.3.3}$$

Now $H(j\omega)$ consists of an infinite series of ascending powers of ω, and is not obtainable from a system described by a finite number of differential equations with constant coefficients (section 3.3). For the case of an electrical circuit this is equivalent to saying that a response function of the type given in equation (4.3.2) is not obtainable from a circuit with a finite number of discrete ('lumped') components; it is in fact characteristic of an ideal transmission line of length vT_S, where v is the wave velocity in the line and is independent of ω. We refer to this as a *distributed-parameter* system, and we can regard it as being the limiting case of a *'lumped-parameter'* system as the number of components approaches infinity. Such a system is described by *partial* differential equations rather than linear equations with constant coefficients. When we need to delay a signal without distortion, for example in an oscilloscope where we wish to display a signal pulse in the middle of a time-base sweep which was itself triggered by the arrival of the pulse, we must make use of a *delay line*, which usually consists of a length of coaxial cable wound on a drum. The quantity T_S defined above is sometimes referred to as the *signal front delay time* in cases where confusion may arise with *group delay time* T_G introduced in section 4.4.

The distortionless system defined by equation (4.3.2) may be called an 'all-pass system' because A does not vary with ω. Other forms of all-pass system have a different form of $\theta(\omega)$ and will be discussed later. It is also classified as a linear-phase system, because θ is proportional to ω. We may say that any system which is not all-pass shows amplitude distortion, and any system which is not linear-phase shows phase distortion. Systems in general (i.e. apart from special cases) show both amplitude and phase distortion, and there is no implication that they are necessarily undesirable.

The designation 'low-pass' is applied to systems in which $A(\omega)$ is more or less constant from $\omega = 0$ up to a cut-off frequency ω_L in the region of which $A(\omega)$ falls more or less sharply towards zero. If the Fourier components of the input function $x(t)$ extend over a range of frequencies above and below ω_L, such a system will act as a filter in that the output will be substantially due to the components of the input below ω_L.

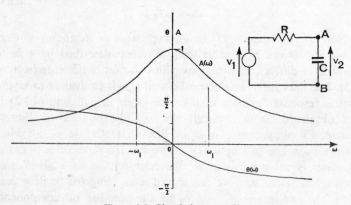

Figure 4.1. Simple low-pass filter.

The simplest low-pass system is represented in figure 4.1, where the output voltage v_2 is related to the input voltage v_1 by means of the frequency-response function

$$H(j\omega) = \omega_1/(j\omega + \omega_1) \text{ where } \omega_1 = 1/RC . \qquad (4.3.4)$$

Clearly, the system function $H(s)$ has a simple pole at the point $-\omega_1$ on the real axis. In terms of magnitude and phase angle we may describe the system by the relations

$$A(\omega) = \omega_1/\sqrt{(\omega^2 + \omega_1{}^2)} \left.\vphantom{\begin{matrix}1\\1\end{matrix}}\right\}$$
$$-\theta(\omega) = \tan^{-1}(\omega/\omega_1) \qquad (4.3.5)$$

The frequency ω_1 may be regarded as being the cut-off frequency of this system; when $\omega = \omega_1$, $A(\omega)$ has fallen by a factor of $1/\sqrt{2}$ from its low-frequency value, and for $\omega \gg \omega_1$, A falls off as $1/\omega$ whilst θ approaches $-\pi/2$.

There are many physical quantities which vary with frequency according to equation (4.3.4), which is often put in the form

$$H(j\omega) = 1/(1 + j\omega T)$$

T being called the *relaxation time*.

A sharper cut-off than that described by equation (4.3.4) is given by a system whose frequency-response function is the square of that in equation (4.3.4). This is

$$H(j\omega) = \omega_1{}^2/(j\omega + \omega_1)^2 \left.\vphantom{\begin{matrix}1\\1\\1\end{matrix}}\right\}$$
$$A(\omega) = \omega_1{}^2/(\omega^2 + \omega_1{}^2) \qquad (4.3.6)$$
$$-\theta(\omega) = 2\tan^{-1}(\omega/\omega_1)$$

Here $A(\omega)$ falls off as $1/\omega^2$ for $\omega \gg \omega_1$ and $\theta(\omega)$ approaches $-\pi$.

Of course, this response cannot be obtained by simply attaching another R–C combination across AB in figure 4.1 and taking the output across the second capacitor, because these additional components will affect the voltage across AB; in fact the resulting system function (assuming the resistors and the capacitors to be equal) will have two poles at $-\frac{1}{2}[3 \pm \sqrt{5}]\omega_1$. We may however realize it by the use of a unity-gain voltage amplifier acting as a 'buffer' between two identical R–C sections, as shown in figure 4.2.

The required characteristics for a buffer amplifier are that it should have a high input impedance so that it has a negligible effect on the output of the first R–C section, and a low output impedance so that its voltage transfer is not substantially affected by the variation of the impedance of the second R–C section with frequency. In some cases these requirements are met by a single transistor in the 'emitter-follower' configuration.

Another important type of low-pass frequency-response is defined by the relation

$$A(\omega) = A_0/\sqrt{[1 + (\omega/\omega_1)^{2n}]}, \; n > 1 . \qquad (4.3.7)$$

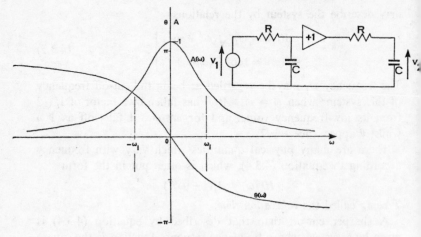

Figure 4.2. System with a frequency – response function $\omega_1{}^2/(j\omega + \omega_1)^2$.

A frequency-response function whose magnitude has this form is called a *Butterworth* response of order n; its special property is that it is 'maximally-flat' in the sense that all the derivatives of $A(\omega)$ up to the $(2n-1)$th are zero at the origin. It is a straightforward problem to derive a Butterworth frequency-response function of any order, fulfilling the conditions of stability and causality and therefore capable of being constructed in practice. Such functions are:

second-order: $H(j\omega) = 1/[\omega^2 - \sqrt{2}j\omega\omega_1 - \omega_1{}^2]$

third-order: $H(j\omega) = 1/[(j\omega + \omega_1)(\omega^2 - j\omega\omega_1 - \omega_1{}^2)]$.

A system having a second-order Butterworth response is shown in figure 4.3. In the terminology of section (3.4), the damping ratio ζ has a value of $1/\sqrt{2}$.

We must now consider *band-pass* systems. Such a system is one where $A(\omega)$ has significant values only in a limited frequency range

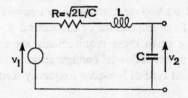

Figure 4.3. Circuit with second-order Butterworth response.

which does not include $\omega = 0$. A band-pass system is said to be *symmetrical* about a *centre frequency* ω_0 (ω_0 being implicitly positive) if its frequency-response obeys the relation

$$H(j\omega) = H^*(2j\omega_0 - \omega_1). \tag{4.3.8}$$

It follows from equation (4.3.8) that $A(\omega)$ is even and $\theta(\omega)$ is odd about ω_0. The response in the negative frequency range is determined by the requirement that $H(-j\omega)$ is the conjugate of $H(j\omega)$, which implies that there must be a corresponding centre frequency at $-\omega_0$.

Now we may regard the frequency-response of a symmetrical band-pass system as being derived from an *equivalent low-pass system* $H_L(j\omega)$ which has been shifted by an amount $\pm\omega_0$ along the frequency axis. Thus

$$H(j\omega) = H_L(j\omega - j\omega_0) + H_L(j\omega + j\omega_0). \tag{4.3.9}$$

Since we have postulated that $A = 0$ for $\omega = 0$, the function $H_L(j\omega)$ must be negligibly small for $|\omega| > \omega_0$. The first term in (4.3.9) is thus effective for positive values of ω, and the second term for negative values; the symmetry relation in (4.3.8) is therefore satisfied by (4.3.9) because of the conjugate symmetry of $H_L(j\omega)$.

The next step is to calculate the impulse response of the system described by (4.3.9). The required inverse Fourier transform is

$$h(t) = (1/2\pi) \int_{-\infty}^{\infty} [H_L(j\omega - j\omega_0) + H_L(j\omega + j\omega_0)] \exp j\omega t \, d\omega \tag{4.3.10}$$

and the evaluation is made easier by a change of variable. Taking the first term in the bracket we obtain,

$$(1/2\pi) \int_{-\infty}^{\infty} H_L(j\omega) \exp [j(\omega + \omega_0)t] \, d\omega = h_L(t) \exp j\omega_0 t. \tag{4.3.11}$$

Here $h_L(t)$ is the impulse-response of the equivalent low-pass system. The second term in (4.3.10) is dealt with in the same way and the total result is:

$$h(t) = 2h_L(t) \cos \omega_0 t \tag{4.3.12}$$

We see that the impulse-response of a symmetrical band-pass system with a centre-frequency ω_0 can be described in terms of the amplitude-modulated carrier discussed in section 4.4. The carrier has a frequency ω_0 and it is amplitude-modulated with an envelope equal to $2 \times$ the impulse response of the equivalent low-pass system.

As an example we may derive a symmetrical band-pass response from the simple low-pass response of equation (4.3.4),

Low-pass: $H_L(j\omega) = 1/(1+j\omega/\omega_1)$, ω_1 positive.

Band-pass: $H(j\omega) = 1/[1+(j\omega-j\omega_0)/\omega_1]+1/[1+(j\omega+j\omega_0)/\omega_1]$.

$$(4.3.13)$$

The system function $H(s)$ has poles at $-\omega_1 \pm j\omega_0$, and a zero at $-\omega_1$. The impulse response of the system can be derived from the impulse response of the equivalent low-pass system; the latter is simply $u(t)\omega_1 \exp -\omega_1 t$, and by the use of equation (4.3.12) we obtain

$$h(t) = 2u(t)\omega_1 \exp -\omega_1 t \cos \omega_0 t. \qquad (4.3.14)$$

An equivalent result was obtained from a different point of view in section 3.4.

When we are considering the response for values of ω in the region of ω_0, we see that if the ratio $|\omega_0|/\omega_1$ is sufficiently high the contribution of the first term in (4.3.13) is much greater than the second and we have a symmetrical band-pass response. The first term (with ω_1 positive) is often used by itself to represent the response of a narrow-band resonant system to frequencies close to the resonance frequency; this is the well-known *Lorentz* function (figure 4.4). It must be borne in mind that this function does not have the required conjugate-symmetry property of a true frequency-response function (its calculated impulse response being complex

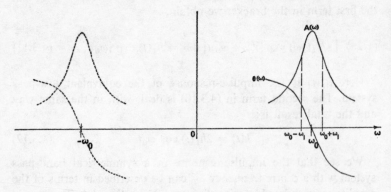

Figure 4.4. The Lorentz function.

rather than real) and furthermore that it is not applicable for values of ω which are far from the resonance region.

When the Lorentz function is used to describe an electrical resonant system, it is usually written

$$H(j\omega) = 1/[1+2jQ(\omega-\omega_0)/\omega_0] \qquad (4.3.15)$$

$$\text{where } Q = \omega_0/2\omega_1 = 1/2\zeta .$$

ζ being the damping ratio discussed in section 3.4. Practically speaking, this formulation is usually sufficiently accurate to describe the response of a second-order resonant system (typically, an L–R–C circuit) in the region of resonance provided that Q exceeds about 10; for lower values of Q the asymmetry becomes more pronounced and it becomes necessary to take account of both the terms in (4.3.13).

In the treatment of resonant systems in other fields of physics the Lorentz function often appears in the form

$$H(j\omega) = 1/[1+j(\omega-\omega_0)T] \qquad (4.3.16)$$

where T, the reciprocal of ω_1, is the amplitude decay time of free oscillations and is called the *relaxation time* of the system.

It is clear from (4.3.13) that, in the symmetrical approximation, the parameter ω_1 defines the value of $|\omega-\omega_0|$ for which the magnitude of $H(j\omega)$ falls to $1/\sqrt{2}$ from its peak value of unity at $\omega = \omega_0$; hence ω_1, ζ, Q, and T are alternative ways of specifying the width of the band-pass function.

4.4 Group delay and signal-front delay

In this section we are concerned with the response of a system to an input of the *amplitude-modulated carrier* type:

$$f_M(t) = f_A(t) \cos \omega_C t . \qquad (4.4.1)$$

This input function is of basic importance in communications. $f_A(t)$ is called the *modulation function* or *envelope function* and in practical cases it is restricted to frequencies well below the carrier frequency ω_C. If $f_A(t)$ is a periodic function with fundamental frequency ω_A, it can be expressed as a Fourier series and $f_M(t)$ has a number of discrete Fourier components at frequencies $(\omega_C \pm n\omega_A)$ where n is, of course, an integer. In general $f_A(t)$ will not be periodic and it will be described by its Fourier transform $F_A(j\omega)$ where

$$F_A(j\omega) = \int_{-\infty}^{\infty} f_A(t) \exp -j\omega t \, dt . \qquad (4.4.2)$$

The Fourier transform of the input function (4.4.1) is then

$$F_M(j\omega) = \int_{-\infty}^{\infty} f_A(t) \cos \omega_C t \exp -j\omega t \, dt = \\ \tfrac{1}{2}F_A(j\omega - j\omega_C) + \tfrac{1}{2}F_A(j\omega + j\omega_C) . \qquad (4.4.3)$$

If the frequency range of the modulation function is limited to frequencies well below ω_C, $F_M(j\omega)$ will consist of two relatively narrow bands in the region of ω_C and $-\omega_C$, as shown in figure 4.5.

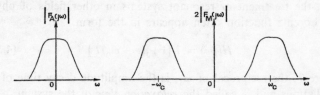

Figure 4.5. Fourier transform of the function $f_M(t) = f_A(t) \cos \omega_C t$.

We shall now assume that, for the system under consideration, the magnitude of $H(j\omega)$ is substantially constant in the frequency range occupied by the input.

We further assume that we can ignore all the derivatives of the phase-angle variation higher than the first, so that for the limited region about $\omega = \omega_C$ we may write

$$\left. \begin{array}{c} A(\omega) = A_C, \ \theta(\omega) = \theta_C + (\omega - \omega_C)T_G \\ \text{where } T_G = [d\theta(\omega)/d\omega] . \end{array} \right\} \qquad (4.4.4)$$

First we shall consider the simplest form of amplitude – modulated input, which is the case of sinusoidal modulation giving two discrete sideband frequencies,

$$x(t) = x_0 \cos \omega_1 t \cos \omega_C t$$

$$= \tfrac{1}{2}x_0 \cos (\omega_C + \omega_1)t + \tfrac{1}{2}x_0 \cos (\omega_C - \omega_1)t \qquad (4.4.5)$$

and if we use this function as the input to the system described by equations (4.4.4) the output $y(t)$ will be

$$y(t) = A_C x_0 \cos \omega_1(t+T_G) \cos (\omega_C t + \theta_C) \qquad (4.4.6)$$

The overall effect of the system is thus equivalent to a shift of the carrier phase by θ_C, accompanied by a shift of the modulation function by a time T_G. Now this time does not depend on ω_1, a fact which is easily understood when we realize that a phase-shift in the modulation function is achieved by a *relative* shift between the phases of the two sideband frequencies. This shift depends on $d\theta(\omega)/d\omega$ and is also proportional to the separation of the two sideband frequencies, that is to ω_1. A phase-shift which is proportional to frequency is expressible as a time delay, which may of course be positive or negative according to the sign of $d\theta(\omega)/d\omega$. A delay of the modulation function is called *group-delay*, and T_G is the *group-delay time*.

On the assumptions we have made, the same group delay time T_G would apply to all the Fourier components of a modulation function $f_A(t)$, and as a result of this we would expect the whole envelope to be delayed by T_G.

It is an interesting exercise to derive this result directly by the use of the Fourier transform. As a first step we must express the frequency-response function in the form

$$H(j\omega) = \begin{cases} A_C \exp j[\theta_C+(\omega-\omega_C)T_G], & \omega \simeq \omega_C \\ A_C \exp j[-\theta_C+(\omega+\omega_C)T_G], & \omega \simeq -\omega_C . \end{cases} \qquad (4.4.7)$$

The input function and its transform, as in equation (4.4.3), are

$$x(t) = f_A(t) \cos \omega_C t$$
$$X(j\omega) = \tfrac{1}{2}F_A(j\omega-j\omega_C)+\tfrac{1}{2}F_A(j\omega+j\omega_C) . \qquad (4.4.8)$$

Now because the input occupies a relatively narrow frequency range, we may state

$$F_A(j\omega) = 0 \text{ for } |\omega| > \omega_C \qquad (4.4.9)$$

so that two of the four terms of the product $Y(j\omega) = X(j\omega)H(j\omega)$ are zero. After some rearrangement we obtain the inverse transform of this function in the form

$$y(t) = f_A(t+T_G) \cos (\omega_C t + \theta_C) . \qquad (4.4.10)$$

Equation (4.4.6) is of course a special case of this result.

Now this phenomenon of group delay, which we have been discussing in relation to the envelope of a modulated carrier, is a property of any linear system possessing a finite value of the co-efficient $d\theta(\omega)/d\omega$ in the region of a specified carrier frequency. It is not in general the same as the signal-front delay defined by equations (4.3.1) and (4.3.2) in relation to a distortionless system (ideal delay line) which is the same for any type of input function because $d\theta(\omega)/d\omega$ is independent of ω. As another example of a system showing signal-front delay we now consider the frequency-response function

$$H(j\omega) = H_1(j\omega) \exp -j\omega T_S$$
$$= A_1(\omega) \exp j\theta_1(\omega) \exp -j\omega T_S \qquad (4.4.11)$$

where H_1 is a rational function of ω. This is equivalent to a system $H_1(j\omega)$ connected through a suitable buffer amplifier to a delay line with delay time T_S. The group-delay time referred to any given carrier frequency can be obtained from the relation

$$T_G = d\theta(\omega)/d\omega = d\theta_1(\omega)/d\omega + T_S . \qquad (4.4.12)$$

Now if we consider the form of the rational function $H_1(j\omega)$ as $\omega \to \infty$ we see that $\theta_1(\omega)$ approaches a constant value and we may write

$$T_S = \lim_{\omega \to \infty} d\theta(\omega)/d\omega . \qquad (4.4.13)$$

This relation may be used as a definition of the signal-front delay time, T_S, for systems of this type.

4.5 Minimum-phase criterion
In section 4.2 we have shown that if $H(j\omega)$ is the transform of a causal function $h(t)$, that is if $H(j\omega)$ is the frequency-response function of a realizable system, then there exists a unique relationship between $H_R(\omega)$ and $H_I(\omega)$ so that if we are given $H_R(\omega)$ we can calculate $H_I(\omega)$ and vice-versa.

We may also express the frequency-response function in the polar form

$$H(j\omega) = A(\omega) \exp j\theta(\omega) .$$

Now in this formulation the phase-angle function $\theta(\omega)$ is not uniquely determined by the magnitude function $A(\omega)$. We can see this by considering the properties of the all-pass system

$$\left.\begin{array}{l} H_{AP}(s) = (s-a)/(s+a) \text{ [a real and positive]} \\[2mm] A_{AP}(\omega) = 1 \\[2mm] -\theta_{AP}(\omega) = 2\tan^{-1}(\omega/a) . \end{array}\right\} \quad (4.5.1)$$

This is a causal and stable system, $H_{AP}(s)$ being a rational function of s with no poles in the right-half s-plane. If the input of such a system were connected (if necessary through a suitable buffer amplifier) to the output of another stable system, the combined system would have the same magnitude function as the original system, but a greater phase-lag in the high-frequency limit.

Generally speaking, we can multiply any stable system function $H(s)$ by the factor $(s-a)/(s+a)$, a being real and positive, to obtain another stable system function with a frequency-response having the same magnitude as before, but with a greater phase-lag.

However, it is not possible in general to *reduce* the phase-lag of a system by means of the factor $(s+a)/(s-a)$, (a real and positive) because in general the introduction of a pole at $+a$ will make the system function into an unstable one. The exception is the case where the original system function has a zero at $+a$, that is a factor $(s-a)$ in the numerator.

We conclude that any stable system function having a zero at z, in the right-half s-plane can be 'converted' into a stable system function having the same amplitude – but smaller phase-lag – by multiplication with the factor $(s+z_1)/(s-z_1)$. This leads to the definition of a *minimum-phase* system function as *one which has neither zeros nor poles in the right-half s-plane*. Furthermore, the minimum-phase function with the same magnitude function as a given non-minimum-phase function can be obtained by the simple process of multiplying by factors of the form $(s+z_n)/(s-z_n)$, where the quantities z_n are those zeros of the non-minimum-phase function which lie in the right-half s-plane.

Conversely, we can regard any non-minimum-phase function as consisting of the product of a minimum-phase function and factors of the form $(s-z_n)/(s+z_n)$, where the complex quantities z_n lie in the right-half s-plane.

It is always true that the driving-point impedance (that is, the relation between voltage and current at the same pair of terminals)

c

of a passive network is a minimum-phase function. On the other hand, the transfer impedance between two pairs of terminals in a passive network may or may not be minimum-phase. The essential difference between these two cases is that, because a passive network always absorbs power from the source, the real part of its driving point impedance is positive at all real frequencies; as we shall show in section 5.7, this property in a system-function is sufficient to establish it as a minimum phase function. The converse proposition is not true – a system whose real part varies in sign along the $j\omega$ axis may or may not be minimum-phase.

CHAPTER 5

INTEGRAL THEOREMS

5.1 Integration of $(s-p)^n$

Before proceeding to establish the integral theorems which are the subject of this chapter, we must remind ourselves of a result which is of great importance in the evaluation of contour integrals. This refers to the integral of the function $(s-p)^n$ along the path Γ shown in figure 5.1; Γ is an arc of a circle of radius r centred at p, with

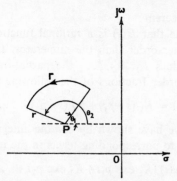

Figure 5.1. Contour for integration of $(s-p)^n$.

length defined by radius vectors making angles θ_1 and θ_2 with the real axis.

For points along Γ we may write

$$(s-p) = r \exp j\theta; \quad ds = jr \exp j\theta \, d\theta \qquad (5.1.1)$$

and the required integral is

$$\int_\Gamma (s-p)^n ds = jr^{n+1} \int_{\theta_1}^{\theta_2} \exp [j(n+1)\theta] \, d\theta$$

$$= \begin{cases} [r^{n+1}/(n+1)] \{\exp [j(n+1)\theta_2] - \exp [j(n+1)\theta_1]\} & \text{for } n \neq -1 \\ j(\theta_2 - \theta_1) & \text{for } n = -1 \end{cases}$$

$$(5.1.2)$$

C*

Now if Γ is a complete circle, $(\theta_2 - \theta_1)$ is equal to 2π and the integral vanishes except for $n = -1$, where it has a value of $2\pi j$.

Now let us consider a function $F(s)$ having a first-order pole of residue K at the point p. By definition we have

$$F(s) \to K/(s-p) \text{ as } s \to p \ . \qquad [(3.3.2)]$$

If we integrate $F(s)$ along the path Γ, we see from (5.1.2) that the integral approaches $jK(\theta_2 - \theta_1)$ as the radius r approaches zero. Thus

$$\lim_{r \to 0} \int_{\Gamma} F(s) \, ds = jK(\theta_2 - \theta_1) \ . \qquad (5.1.3)$$

Now if Γ is a semicircle, the limiting value of the integral is $\pi j K$; this result will be used in this chapter for the evaluation of some important contour integrals.

5.2 Initial-value theorem

Let us first assume that $H(s)$ is a rational function of s with denominator of higher order than the numerator. If $H(s)$ has poles $p_1, p_2, \ldots$ of residues $K_1, K_2, \ldots$, the partial-fraction expansion will contain first-order fractions of the following type:

$$H(s) = K_1/(s-p_1) + K_2/(s-p_2) + \ldots \ .$$

In section 3.3 we have shown by contour integration that these fractions give rise to corresponding terms in the impulse response:

$$h(t) = u(t) \, (K_1 \exp p_1 t + K_2 \exp p_2 t + \ldots) \ . \qquad [(3.3.14)]$$

Now putting $t = 0$ in (3.3.14) and remembering that $u(0+) = 1$, we have

$$h(0+) = (K_1 + K_2 + \ldots) \ . \qquad (5.2.1)$$

The partial-fraction expansion of $H(s)$ may also contain fractions of the form $K_a/(s-p_a)^2$ derived from second-order poles. Equation (3.3.16) shows that such fractions contribute terms of the type $u(t)K_a t \exp p_a t$ to the impulse response. These terms have zero value for $t = 0+$, as have terms in $t^2 \exp p_a t$, etc. corresponding to poles of third and higher order. We conclude that

$$h(0+) = \mathscr{K} \qquad (5.2.2)$$

where the symbol $\mathscr{K}$ represents the *sum of the residues* of $H(s)$ at all its poles.

Now if we consider the behaviour of the terms in the partial fraction expansion as $|s|$ increases we see that if $\mathcal{K}$ is non-zero

$$H(s) \rightarrow \mathcal{K}/s \text{ as } |s| \rightarrow \infty. \tag{5.2.3}$$

In particular, if we let s increase along the $j\omega$ axis we have

$$\left.\begin{array}{c} \omega H_I(\omega) \rightarrow -\mathcal{K} \\[2mm] \omega H_R(\omega) \rightarrow 0 \end{array}\right\} \text{ as } \omega \rightarrow \infty \tag{5.2.4}$$

It follows that we can rewrite equation (5.2.2) in the alternative form

$$h(0+) = \lim_{\omega \to \infty} [-\omega H_I(\omega)]. \tag{5.2.5}$$

It is important to notice that the form of the impulse response near $t = 0$ is determined by the form of $H_I(\omega)$ in the high-frequency limit. If the system function is such that $H_I(\omega)$ falls off with increasing frequency faster than $1/\omega$, then $h(0+)$ is zero.

In cases where the numerator of $H(s)$ is of the same order as the denominator, a stage of long division is necessary before partial-fraction expansion can proceed and the function takes the form

$$H(s) = K_0 + \text{(partial fractions)}. \tag{5.2.6}$$

The output $y(t)$ therefore includes a term $K_0 x(t)$, where $x(t)$ is the input function, and – in particular – the impulse response itself contains an impulse, and $h(0+)$ is infinite.

Where the numerator of the system function is of higher order than the denominator, the impulse response contains derivatives of impulses. Such functions will not concern us here because, as discussed in section 2.2, they are not usually the most satisfactory representation of a physical system.

In the following sections we shall be assuming that $H(s) \rightarrow 0$ as $|s| \rightarrow \infty$, and also that $H(s)$ is a rational function of s. This latter assumption is made in order to simplify the discussion, and must not be regarded as defining a necessary condition for the validity of the results.

5.3 Real-part integral theorem

We shall now consider the result of integrating $H(s)$ along the contour shown in figure 5.2.

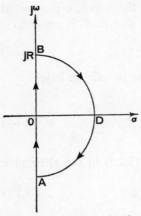

Figure 5.2. Contour for integration of $H(s)$.

As R increases towards infinity, we see that for points on the semicircle BDA the function $H(s)$ approaches $\mathscr{K}/s$ [as in equation (5.2.3)], where $\mathscr{K}$ is the sum of the residues of $H(s)$, if $\mathscr{K}$ is non-zero. Using equation (5.1.2) we may therefore write

$$\lim_{R\to\infty} \int_{\text{BDA}} H(s)\,\mathrm{d}s = -\mathrm{j}\pi\mathscr{K}\,. \qquad (5.3.1)$$

Also, since we are considering a stable system, $H(s)$ can have no poles in the right-half s–plane and is therefore analytic within the contour; thus

$$\int_{\text{AOBDA}} H(s)\,\mathrm{d}s = 0 = \int_{\text{BDA}} H(s)\,\mathrm{d}s + \int_{\text{AOB}} H(s)\,\mathrm{d}s\,. \qquad (5.3.2)$$

Taking this last result in conjunction with equations (5.3.1) and (5.2.2) we obtain

$$\lim_{R\to\infty} \int_{\text{AOB}} H(s)\,\mathrm{d}s = \mathrm{j}\pi h(0+)\,. \qquad (5.3.3)$$

Now this integral is along the $\mathrm{j}\omega$ axis so that we may rewrite it in the real-frequency form

$$\int_{-\infty}^{\infty} H(\mathrm{j}\omega)\,\mathrm{d}\omega = \pi h(0+)\,. \qquad (5.3.4)$$

Now since $H(\mathrm{j}\omega)$ is the Fourier transform of a real function $h(t)$, its real part $H_R(\omega)$ and its imaginary part $H_I(\omega)$ are even and odd functions respectively – see equations (4.1.2) and (4.1.3). Equation (5.3.4) therefore becomes, in conjunction with (5.2.5)

$$\int_{-\infty}^{\infty} H_R(\omega)\,\mathrm{d}\omega = 2\int_{0}^{\infty} H_R(\omega)\,\mathrm{d}\omega = \pi h(0+)$$
$$= \pi \lim_{\omega\to\infty} [-\omega H_I(\omega)]\,. \qquad (5.3.5)$$

This result is known as the *real-part integral theorem* and it shows that the area under the $H_R(\omega)$ curve is determined, like

$h(0+)$, by the form of $H_I(\omega)$ at the high-frequency limit. Notice that if we substitute $t = 0$ in the inversion integral

$$h(t) = (1/2\pi) \int_{-\infty}^{\infty} H(j\omega) \exp j\omega t \; d\omega \qquad [(4.1.6)]$$

we arrive at a value of $h(0)$ which is equal to $\frac{1}{2}h(0+)$. This value is the average of $h(0+)$ and $h(0-)$, the latter quantity being of course zero. It is a general property of the inverse Fourier transform to give an average value in this way at points where the required function has a discontinuity.

The result expressed by equation (5.3.3) can equally well be obtained by integration along a contour completed by a semicircle in the left-half s-plane. Here the integral along the semicircle has the opposite sign from that given in equation (5.3.1) because the integration is performed in an anticlockwise rather than a clockwise direction. However, this choice of contour encloses all the poles of $H(s)$ so that its value as $R \to \infty$ is $2\pi j \mathcal{K}$. Thus, the same result is obtained for the real-part integral.

In many practical systems the integral $\int H_R(\omega) \; d\omega$ is important as a measure of the *gain-bandwidth product*. The real-part integral theorem shows that if we modify a system so as to change $H_R(\omega)$ (for example by applying feedback), the gain-bandwidth product remains constant provided that the modification has no effect on the value of $H_I(\omega)$ in the high-frequency limit. Another interesting point is that, for systems in which $h(0+)$ is zero, the real-part integral vanishes – this implies that $H_R(\omega)$ takes both positive and negative values.

5.4 Integral of $[H(j\omega)]^2$

The function $[H(s)]^2$ has poles and zeros at the same points as $H(s)$. From equation (5.2.3) we have, if $\mathcal{K}$ is non-zero,

$$[H(s)]^2 \to \mathcal{K}^2/s^2 \text{ as } |s| \to \infty \qquad (5.4.1)$$

and, as a result, the integral along the semicircular path BDA shown in figure 5.2 vanishes as $R \to \infty$. Since, of course, the contour encloses no poles, the integral around the whole contour AOBDA is also zero. We could equally well have drawn the semicircle in the left-hand s-plane and the integral would still have been zero because, as we can see from (5.4.1), the sum of the residues at the poles of $[H(s)]^2$ is zero. It follows from this that the integral along AOB approaches zero as $R \to \infty$, thus

$$\int_{-\infty}^{\infty} [H(j\omega)]^2 d\omega = 0 . \qquad (5.4.2)$$

In terms of the real and imaginary parts of $H(j\omega)$ this becomes

$$\int_{-\infty}^{\infty} [H_R(\omega)]^2 d\omega - \int_{-\infty}^{\infty} [H_I(\omega)]^2 d\omega + 2j \int_{-\infty}^{\infty} H_R(\omega)H_I(\omega)\, d\omega = 0 . \qquad (5.4.3)$$

Now the last term in equation (5.4.3) vanishes because $H_R(\omega)$ is an even function and $H_I(\omega)$ an odd function of ω, so that we may write

$$\int_{0}^{\infty} [H_R(\omega)]^2 d\omega = \int_{0}^{\infty} [H_I(\omega)]^2 d\omega . \qquad (5.4.4)$$

This result shows one aspect of the relationship between the real part and the imaginary part of a causal system function.

5.5 The Kramers-Kronig relations

In section 4.2 it was pointed out that because $H_R(j\omega)$ and $H_I(j\omega)$ are the Fourier cosine transform and the Fourier sine transform respectively of a causal function of time, it is possible in principle to deduce $H_R(j\omega)$ from $H_I(j\omega)$ and vice-versa. The formulae given in equations (4.2.5) and (4.2.7) are not in a very convenient form and we shall now use contour integration to derive some more useful relationships.

We shall find that we can obtain the required results by integrating the function $H(s)/(s-j\omega_1)$ around the contour illustrated in figure 5.3.

Now since $H(s)$ is a rational function which approaches zero as $|s| \to \infty$, then $H(s)/(s-j\omega_1)$ approaches zero at least as fast as $1/s^2$ and its integral along BDA in figure 5.3 vanishes as $R \to \infty$. Also $H(s)/(s-j\omega_1)$ has poles and zeros at the same points as $H(s)$ with an additional pole at $j\omega_1$ on the imagin-

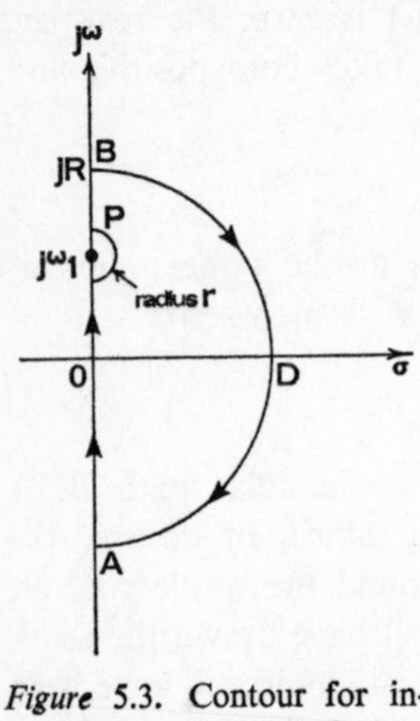

Figure 5.3. Contour for integration of $H(s)/(s-j\omega_1)$.

ary axis. The contour in figure 5.3 encloses no poles of $H(s)$, being in the right-half s-plane, and makes a small semicircular detour around $j\omega_1$, so that the integral around AOPBDA is zero. We may therefore write

$$\int_{\text{AOPB}} [H(s)/(s-j\omega_1)]ds \to 0 \text{ as } R \to \infty . \qquad (5.5.1)$$

Now the residue of $H(s)/(s-j\omega_1)$ at the pole at $j\omega_1$ is $H(j\omega_1)$, and by equation (5.1.3), the integral along the small semicircular path P approaches $\pi j H(j\omega_1)$ as $r \to 0$. Apart from P, the integration AOPB is entirely along the $j\omega$ axis and (remembering that $ds = jd\omega$) we may put equation (5.5.1) in the form

$$\lim_{\substack{r\to 0 \\ R\to\infty}} \left\{ \begin{array}{l} \displaystyle\int_{-\infty}^{\omega_1-r} [H(j\omega)/(\omega-\omega_1)]d\omega \\[2ex] + \displaystyle\int_{\omega_1+r}^{\infty} [H(j\omega)/(\omega-\omega_1)]d\omega + j\pi H(j\omega_1) = 0 . \end{array} \right. \qquad (5.5.2)$$

Now the sum of the first two terms in (5.5.2) is, by definition, the *Cauchy principal value* of the integral along the $j\omega$ axis. We shall denote the principal value by a cross through the integral sign and write

$$\fint_{-\infty}^{\infty} [H(j\omega)/(\omega-\omega_1)]d\omega + j\pi H(j\omega_1) = 0 . \qquad (5.5.3)$$

Equating real and imaginary parts in this last result we reach the conclusion

$$\left. \begin{array}{l} (1/\pi) \displaystyle\fint_{-\infty}^{\infty} [H_R(\omega)/(\omega-\omega_1)]d\omega = H_I(\omega_1) \\[3ex] -(1/\pi) \displaystyle\fint_{-\infty}^{\infty} [H_I(\omega)/(\omega-\omega_1)]d\omega = H_R(\omega_1) \end{array} \right\} . \qquad (5.5.4)$$

These results provide a convenient method of calculating H_R from H_I and vice-versa.

We can change the lower limit of integration from $-\infty$ to zero if we take into account the fact that $H_R(\omega)$ is an even function and

$H_1(\omega)$ an odd function of ω. We then obtain the expressions

$$\left.\begin{array}{l} H_1(\omega_1) = -(2\omega_1/\pi) \displaystyle\oint_0^\infty [H_R(\omega)/(\omega_1{}^2 - \omega^2)]d\omega \\[4mm] H_R(\omega_1) = (2/\pi) \displaystyle\oint_0^\infty [\omega H_1(\omega)/(\omega_1{}^2 - \omega^2)]d\omega \end{array}\right\} \qquad (5.5.5)$$

which are often called the *Kramers-Kronig* relations.

If we substitute $\omega_1 = 0$ in the second of equations (5.5.5) we find that

$$H_R(0) = -(2/\pi) \int_0^\infty [H_1(\omega)/\omega]d\omega . \qquad (5.5.6)$$

This result is called the *imaginary-part integral theorem*, or the *reactance integral theorem*.

5.6 Relationship between gain and phase

In section 4.7 we have established the fact that when a system function is expressed in the polar form

$$H(j\omega) = A(\omega) \exp j\theta(\omega)$$

there is no unique relationship between $A(\omega)$ and $\theta(\omega)$. However, we have defined a *minimum-phase* function as one which has neither zeros nor poles in the right-half s–plane and there is only one minimum-phase function corresponding to a given $A(\omega)$. We shall now establish the relation between $A(\omega)$ and $\theta(\omega)$ for a minimum-phase function.

We shall find it convenient to define a new quantity

$$a(\omega) = \log [A(\omega)] \qquad (5.6.1)$$

so that

$$\log [H(j\omega)] = a(\omega) + j\theta(\omega) . \qquad (5.6.2)$$

The quantity $a(\omega)$ may be called the *logarithmic gain* of the system, and in practice it is often much more convenient to work with than $A(\omega)$. However, in the present discussion its importance lies in the fact that $a(\omega)$ and $\theta(\omega)$ are the real and imaginary parts of $\log [H(j\omega)]$. We shall establish the relationship between them for a minimum-phase function by a contour integration involving

$\log [H(s)]$; in this operation the special feature of the minimum-phase function is that not only $H(s)$ but also $\log [H(s)]$ is analytic at all points in the right-half s–plane.

To ensure that the integral will converge we must examine the behaviour of $\log [H(s)]$ as $|s|$ approaches infinity.

$$\left. \begin{array}{ll} \text{If} & H(s) \to K/s^n \\[2mm] \text{then} & \log [H(s)] \to \log K - n \log s \end{array} \right\} \text{ as } |s| \to \infty . \qquad (5.6.3)$$

Accordingly we set up the function

$$F_1(s) = \log [H(s)]/(s^2 + \omega_1^2) \qquad (5.6.4)$$

which is analytic in the right-half s–plane and whose integral along the path BDA in figure 5.4 vanishes as $R \to \infty$.

The contour makes small semicircular detours P and Q of radius r around the poles of $F_1(s)$ at $\pm j\omega_1$. Writing $F_1(s)$ in the form $\log [H(s)]/(s+j\omega_1) (s-j\omega_1)$ we see that the residue at $+j\omega_1$ is

$$\log [H(j\omega_1)]/2j\omega_1.$$

Similarly the residue at $-j\omega_1$ is

$$-\log [H(-j\omega_1)]/2j\omega_1.$$

In the same way that we derived equation (5.5.3) we obtain by integration around the contour APQBDA the result

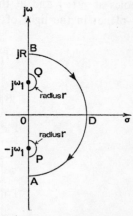

Figure 5.4. Integration of log $[H(s)]/(s^2+\omega_1^2)$.

$$j \oint_{-\infty}^{\infty} [\log [H(j\omega)]/(\omega_1^2 - \omega^2)]d\omega$$

$$+(\pi/2\omega_1)\{\log [H(j\omega_1)] - \log [H(-j\omega_1)]\} = 0 . \qquad (5.6.5)$$

Now we can make use of the definition (5.6.2) in conjunction with the fact that a is an even function and θ an odd function of ω, to obtain the relation

$$\theta(\omega_1) = (\omega_1/\pi) \oint_{-\infty}^{\infty} [a(\omega)/(\omega^2 - \omega_1^2)]d\omega . \qquad (5.6.6)$$

Because $a(\omega)$ is an even function, we can equally well put this in the form

$$\theta(\omega_1) = (2\omega_1/\pi) \oint_0^\infty [a(\omega)/(\omega^2 - \omega_1^2)]d\omega . \qquad (5.6.7)$$

Now the integration of $F_1(s)$, the function defined in equation (5.6.4), has led to an expression for $\theta(\omega)$ in terms of $a(\omega)$. In order to establish the corresponding expression for $a(\omega)$ in terms of $\theta(\omega)$ we make use of the function

$$F_2(s) = \log [H(s)]/s(s^2 + \omega_1^2) \qquad (5.6.8)$$

whose denominator is an odd function of s. This function has poles at $\pm j\omega_1$ and at the origin and is analytic in the right-half s-plane; in the light of the preceding discussion we can see that

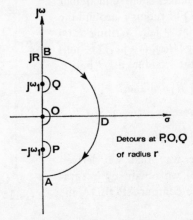

Figure 5.5. Integration of $\log [H(s)]/s(s^2 + \omega_1^2)$.

its integral along the semicircle BDA in figure 5.5 vanishes as $R \to \infty$. The residues at its poles are

$$\text{poles} \begin{cases} +j\omega_1 & -\log [H(j\omega_1)]/2\omega_1^2 \\ 0 & \log [H(0)]/\omega_1^2 \\ -j\omega_1 & -\log [H(-j\omega_1)]/2\omega_1^2 \end{cases} \text{residues} \qquad (5.6.9)$$

where we have assumed that $H(s)$ does not have a zero at the origin, so that $\log [H(0)]$ is finite.

Integrating around the contour APOQBDA in figure 5.5 we obtain

$$\oint_{-\infty}^{\infty} [\log [H(j\omega)]/\omega(\omega_1{}^2 - \omega^2)]d\omega - (j\pi/2\omega_1{}^2)\{\log [H(j\omega_1)]$$

$$+\log [H(-j\omega_1)] - 2 \log [H(0)]\} = 0 . \qquad (5.6.10)$$

Equating real and imaginary parts, and subsequently changing the lower limit of integration from $-\infty$ to 0, we arrive at the final result

$$a(\omega_1) = a(0) + (2\omega_1{}^2/\pi) \int_0^{\infty} [\theta(\omega)/\omega(\omega_1{}^2 - \omega^2)]d\omega . \qquad (5.6.11)$$

Although, according to equation (5.6.7), we can determine $\theta(\omega)$ for a minimum-phase function once $a(\omega)$ is known, we see from equation (5.6.11) that a knowledge of $\theta(\omega)$ is not by itself sufficient for the determination of $a(\omega)$ because the zero-frequency gain $a(0)$ must be known as well. The physical meaning of this becomes clear when we remember that the additive constant $a(0)$ in $a(\omega)$ is equivalent to a multiplicative constant exp $a(0)$ in $A(\omega)$; however much we know about the variation of phase-angle with frequency we are obviously not able to deduce the numerical value of the amplitude without further information. It is sufficient to know the value of $a(\omega)$ at *any* specified frequency because $a(0)$ can then be deduced from equation (5.6.11).

5.7 Poles and zeros within a contour
The last integral theorem to be discussed here is a very important result obtained by integration of the function

$$[dH(s)/ds]/H(s) \qquad (5.7.1)$$

around a closed contour.

If $H(s)$ has a pole or zero at a point s_1 we may write

$$H(s) = (s - s_1)^n G(s) \qquad (5.7.2)$$

where $G(s)$ is analytic and non-zero at s_1. The index n is positive for a zero, negative for a pole, and $|n|$ is the order of the zero or pole. We now obtain the relations

$$dH(s)/ds = n(s - s_1)^{n-1} G(s) + (s - s_1)^n dG(s)/ds \qquad (5.7.3)$$

$$[dH(s)/ds]/H(s) = n/(s - s_1) + [dG(s)/ds]/G(s) . \qquad (5.7.4)$$

Our function, therefore, has a pole of residue $+n$ at each nth order zero of $H(s)$, and a pole of residue $-n$ at each nth order pole of $H(s)$. The integral of the function (5.7.1) taken anticlockwise around any closed contour is therefore equal to $2\pi j \mathcal{N}$ where $\mathcal{N}$ is the 'weighted total' of the poles and zeros within the contour, with each zero being assigned a positive number equal to its order, and each pole being assigned a negative number which is numerically equal to its order.

Now we may express this function in the alternative form

$$[dH(s)/ds]/H(s) = (d/ds)[\log [H(s)]]$$

$$= (d/ds)[a(s)+j\theta(s)] \qquad (5.7.5)$$

so that the integral around any closed contour C is simply the difference between the initial and final values of $[a(s)+j\theta(s)]$. Thus

$$\underset{C}{\Delta}[a(s)+j\theta(s)] = 2\pi j \mathcal{N} \qquad (5.7.6)$$

where the symbol $\underset{C}{\Delta}$ denotes the difference between the initial and final values after integration around the contour C. The meaning of this result is illustrated in figure 5.6 where the variation of some function $H(s)$ is plotted on a polar diagram for values of s along a contour C.

Clearly the quantity $a(s)$ must return to its original value after an integration from the point X, around the contour and back to

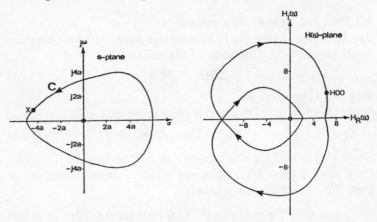

Figure 5.6. Polar plot of a variation of a function $H(s) = 100a^2/(s + a)^2$ along a contour C.

the point X. Thus $\underset{c}{\Delta}[a(s)]$ is zero, a result we can obtain from equation (5.7.6) by equating real parts. On the other hand, $\theta(s)$ can be increased or reduced by an integral multiple of 2π without any effect on the value of $H(s)$. In figure 5.6 we see that, since the locus of $H(s)$ encloses the origin, the integration will have the effect of increasing $\theta(s)$ by 2π; in general $\underset{c}{\Delta}[\theta(s)]$ may be any integral multiple of 2π depending on how many times the locus of $H(s)$ encircles the origin as the contour is traversed. Equating imaginary parts in equation (5.7.6) we obtain

$$\underset{c}{\Delta}[\theta(s)] = 2\pi \mathcal{N} \qquad (5.7.7)$$

which may be stated as follows: *the number of 'zero encirclements' performed by $H(s)$ as s traverses a given contour is equal to the weighted total of the poles and zeros of $H(s)$ within the contour in question.*

One important conclusion which can be drawn from (5.7.7) is the fact, referred to in section 4.7, that the driving-point impedance $Z(s)$ of a passive network is a minimum-phase function. This property arises from the fact that the network must absorb energy from the input source at all real frequencies, so that the real part of $Z(s)$ is positive at all points on the $j\omega$ axis. The real-part integral (equation (5.3.5)) and, consequently, $h(0+)$ are positive quantities and we have from equations (5.2.2.) and (5.2.3)

$$Z(s) \rightarrow h(0+)/s \text{ as } |s| \rightarrow \infty$$

$$[h(0+) \text{ positive}]. \qquad (5.7.8)$$

Now let us consider the values which $Z(s)$ takes along the semi-circular path from $+jR$ to $-jR$ in figure 5.2.

Everywhere along this path the real-part of s is positive, consequently from (5.7.8) the real part of $Z(s)$ is positive. Thus at all points along the contour the real part of $Z(s)$ is positive, so the corresponding locus in the $Z(s)$ plane does not encircle zero. This shows that $\mathcal{N} = 0$ for the whole of the right-half s–plane. Now we already know that $Z(s)$ has no poles in the right-half s–plane; it therefore has no zeros either in this region and must be a minimum-phase function.

In chapter 6 we shall be making use of equation (5.7.7) to establish the Nyquist criterion for the stability of a feedback system.

CHAPTER 6

NEGATIVE FEEDBACK

6.1 The concept of negative feedback

The concept of negative feedback is inextricably bound up with the principle of *control*. A *control mechanism*, or *servomechanism*, has the function of making a physical quantity p approximate to a 'required' value p_R.

The principle of operation of a control system may be illustrated by the following example.

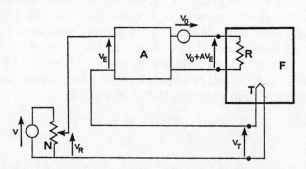

Figure 6.1. A temperature control system.

In figure 6.1 there is shown a diagram of a system designed to control the temperature of a furnace F at a temperature T_R. The furnace incorporates a thermocouple T, whose output voltage V_T is a measure of the temperature of the furnace. The required value of the temperature is 'inserted' into the apparatus by means of a potential-divider network N, adjusted so that its output voltage V_R is equal to the value that V_T would have if the furnace were actually at the required temperature T_R. The quantity $V_E = (V_R - V_T)$ is called the *error voltage* and is zero when the temperature of the furnace is T_R. The error voltage is applied to the input terminals of a voltage amplifier whose system function can be assumed to approximate to a real number A over the frequency range being considered, and the arrangement is such that the voltage across the heating element R is equal to $(V_0 + AV_E)$, where the constant V_0 is large enough to prevent a reversal of sign. Now, if the apparatus is suitably constructed, the furnace will reach an equilibrium temperature very close to T_R. The equilibrium is maintained because if, for any reason, the temperature instantaneously falls, the error voltage increases and the power in R goes up. Similarly an increase in temperature leads to a reduction in the heating power. The accuracy of the control will obviously depend on the amplification factor A; but if A were made too high the effect of the thermal time-constants would cause the control to 'overshoot' in each direction in turn, so that it never reached equilibrium but oscillated from one extreme to the other.

Assuming that the apparatus can be made stable with a satisfactorily high value of A, we will have achieved an arrangement which will automatically adjust the voltage across R to the required value even though we may not know what the required voltage is. As external conditions (for instance the ambient temperature) change, the voltage will change accordingly to keep the temperature at the required value.

This system can be regarded as being typical of control systems in general. The aim is to obtain a constant and well-defined performance from an apparatus some of whose parameters are only approximately known, or which are likely to change from time to time.

The essential features are

(a) the provision of an *error signal* representing the departure of the controlled quantity from its required value, and

(b) *feedback* of the error signal through an error amplifier in such a way as to cause a change of the controlled quantity in the required direction.

6.2 The feedback equation

A linear feedback system has the general form illustrated in figure 6.2. The output quantity, which we represent by its transform $Y(s)$, is derived from system A with system function $A(s)$. The input to system A is the sum of the input quantity $X(s)$ and the quantity $\beta(s)Y(s)$, derived from the output of A via the feedback path β. We therefore have the relations

$$Y(s) = A(s)[X(s)+\beta(s)Y(s)]$$
$$= A_C(s)X(s)$$
where $\quad A_C(s) = A(s)/[1-A(s)\beta(s)]$

(6.2.1)

The combined arrangement of system A, system β, and the adder thus constitutes a linear system with the new system function $A_C(s)$. This system will be stable if $A_C(s)$ has no poles in the right-half s–plane. Clearly the feedback system may be unstable even if $A(s)$ and $\beta(s)$ are themselves stable.

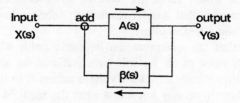

Figure 6.2. A linear feedback system.

The quantities $A(s)$, $A_C(s)$, and $A(s)\beta(s)$ are known respectively as the *open-loop* system function, the *closed-loop* system function, and the *loop gain*. If the system is stable and, in a given frequency range, the magnitude of $[1-A(s)\beta(s)]$ is greater than unity (so that the closed-loop gain is less than the open-loop gain) the feedback path $\beta(s)$ is said to provide *negative feedback*. Similarly if the magnitude of $[1-A(s)\beta(s)]$ is less than unity, $\beta(s)$ provides *positive feedback*. Obviously a given system may show negative feedback in one frequency range and positive feedback in another.

In a given physical situation it is often quite an arbitrary decision to make an analysis in terms of separate A and β systems rather than in terms of a single system from the start. The feedback representation is normally used when one wishes to investigate the effect of varying one branch while keeping the other unchanged.

We must now discuss the relation between the linear feedback system of figure 6.2 and the control system discussed in section 6.1 which arises in the following circumstances. Suppose that we wish to construct a linear system with system function $R(s)$, given that we are able to construct a system A which is capable of delivering the required output and whose system function $A(s)$ is known to be much greater in magnitude than $R(s)$ over the frequency range of interest. Within these limitations $A(s)$ may not be known with certainty – perhaps because it varies substantially with ambient temperature. Suppose also that we can construct a feedback path whose system function $\beta(s)$ is equal to the negative reciprocal of $R(s)$ over the frequency range of interest (it follows that $|A(s)\beta(s)| \gg 1$). We set up the feedback system shown in figure 6.2. If it is stable, it has a system function which approximates to the required function. This is shown in equations (6.2.2).

$$A_C(s) = A(s)/[1 - A(s)\beta(s)]$$
$$\simeq -1/\beta(s) \text{ if } A_C(s) \text{ is stable and if } |A(s)\beta(s)| \gg| . \qquad (6.2.2)$$

In control-system terminology we may say that the β system senses the output and performs on it an operation which is the negative inverse of the required system function; if the whole system were operating in the required fashion, the output of the β system would be equal to the negative of the input function. The adder S performs the operation of comparing the feedback quantity $\beta(s)Y(s)$ with the input $X(s)$, and the difference is the *error* which constitutes the input to system A. The greater the magnitude of $A(s)$ over the frequency range, then the less is the error input which is required to maintain the required output $Y(s)$. The overall system is comparatively insensitive to variations in $A(s)$ (provided of course that the system remains stable), because these cause proportionate variations in the error rather than in the controlled quantity; the effect of such changes can be directly calculated by substitution in the exact expression (6.2.1). A more detailed treatment of feedback in amplifiers is given by the author in *Principles of Linear Circuits*, (Chapman and Hall, 1966).

When we talk of a *negative feedback system* we are usually referring to a system whose performance, under the conditions in which we intend to operate it, is dominated by the effect of an intentionally introduced feedback path β. This implies that the magnitude of the loop-gain $A(s)\beta(s)$, over the frequency range of interest, is greater than, say, 10; however, in many practical cases it is enormously higher. Such a system may reasonably be discussed in terms of 'required output' and 'error' provided that one has some degree of insight into the intentions of the person who constructed it.

6.3 Non-linear distortion

In the previous section we have discussed negative-feedback systems in terms of linear system functions, and we shall now proceed to the case where the system is not perfectly linear. By far the most important reason for our constructing negative-feedback systems is the way in which the overall linearity can be made to depend on the feedback path $\beta(s)$ rather than on the forward path $A(s)$. For example, in the case of the temperature-control system of figure 6.1, it is obvious that the relation between the input voltage of the amplifier and the resulting temperature of the furnace will be a highly non-linear one; however, provided that the temperature-sensing element gives a voltage output which is linearly related to the temperature, and provided that the amplifier has sufficient gain, the equilibrium temperature of the furnace will be a linear function of the input voltage V_R.

In the generalized diagram of figure 6.2, the simplest way of investigating the effect of departures from linearity in system A is to consider a purely sinusoidal input, which we shall represent for convenience by a single complex-exponential term. System A can then be characterized in the following way:

System A alone (open-loop response):

$$\text{input } x(t) = x_1 \exp j\omega_1 t \quad (x_1 \text{ real})$$

$$\text{output } y(t) = x_1 A(j\omega_1) \left[\exp j\omega_1 t + D_{O1} \exp j\omega_1 t \right.$$
$$\left. + D_{O2} \exp 2j\omega_1 t + D_{O3} \exp 3j\omega_1 t + \ldots \right]. \quad (6.3.1)$$

The complex coefficients D_{O1}, D_{O2}, $D_{O3} \ldots$ are called the *open-loop distortion coefficients* and are functions of x_1 and ω_1. Throughout this discussion we shall assume that the magnitudes of the distortion coefficients are small compared with unity, say less than 0.1, under the conditions being considered.

We shall now analyze the effect of negative feedback on the non-linear distortion, with the assumption that the feedback path β is perfectly linear. We must assume that after the introduction of the feedback we have increased the sinusoidal input by such a factor that the first (linear) term in the output is the same as before; it follows from equation (6.2.1) that the required factor is $[1 - A(j\omega_1)\beta(j\omega_1)]$. We can now make use of the distortion coefficients of equation (6.3.1) in calculating the distortion, taking into account the effect of the feedback loop.

Let us consider the nth harmonic frequency. The component $y_n(t)$ of the output at this frequency is the sum of the term $x_1 A(j\omega_1) D_{On} \exp jn\omega_1 t$ and a term generated by the effect of the feedback. Since the distortion coefficients are assumed to be relatively small, we can ignore the non-linearity of the A system when calculating the feedback term and express it to a first-order approximation as $A(jn\omega_1)\beta(jn\omega_1)y_n(t)$. We then have

$$y_n(t) = x_1 A(j\omega_1) D_{On} \exp jn\omega_1 t + A(jn\omega_1)\beta(jn\omega_1)y_n(t)$$

$$= \frac{x_1 A(j\omega_1) D_{On} \exp jn\omega_1 t}{1 - A(jn\omega_1)\beta(jn\omega_1)} \qquad (6.3.2)$$

We may express this result in the form

$$D_{Cn} = D_{On}/[1 - A(jn\omega_1)\beta(jn\omega_1)] \qquad (6.3.3)$$

Here D_{Cn} and D_{On} are the closed-loop and the open-loop nth-harmonic distortion coefficients referred to a given input frequency and output level. The effect of the feedback is to change the harmonic content by a factor which is the familiar feedback ratio evaluated at the frequency of the harmonic in question.

Where the input is not a simple sinusoid, the non-linearity of system A leads to *intermodulation products* between the various Fourier components of the input, at the sum and difference frequencies, in addition to the harmonics shown in (6.3.1). An extension of the argument given above shows that for a given output level, the effect of feedback is to change the total distortion components in the region of any frequency ω by a factor $1/[1 - A(\omega)\beta(\omega)]$, compared with the corresponding open-loop case.

As an example of the application of this principle we shall consider the way in which an electronic engineer designs a broad-band audio-frequency amplifier to give, say, 10 watt output in a given frequency range with less than 0·2 per cent non-linear distortion. At this power level, because of heat-dissipation problems, it is normally a matter of importance to obtain as high an efficiency as possible, and the first step is to design an amplifier which will deliver the required power at high efficiency, keeping the linearity as a secondary consideration. In a typical case this amplifier may turn out to have, say, 20 per cent non-linear distortion at the required maximum output level in the specified frequency range. The circuit is arranged to have (over a frequency range which includes the relevant harmonics) a gain 100 times greater than that required from the completed system, a consideration which normally has no bearing on the question of the linearity or the efficiency because the gain will be determined by low-level transistors incorporated in the early stages of the circuit. The final step is to introduce a feedback path β which reduces the gain by a factor of 100 to the required level, and at the same time reduces the non-linear distortion from 20 per cent to 0·2 per cent under the specified output conditions. The feedback path is a network of resistors and capacitors and therefore 'perfectly' linear – that is with an immeasurably small departure from linearity. In a simple case it may consist of a single resistor, but it will normally be more complicated because of the necessity both to 'shape' the closed-loop system function to the required form and to ensure that the system is stable.

It is interesting to reflect that the completed system could be analyzed by applying Kirchhoff's laws to the whole circuit, making use of the known voltage-current characteristics of all the linear components (resistors and capacitors) and non-linear components (transistors and diodes), and without any reference to the principle of negative feedback. On the other hand it is almost impossible to imagine how the circuit could have been arrived at by the designer without recourse to feedback principles.

6.4 Second-order feedback system

An important special case of negative feedback arises when the forward path $A(s)$ has two real negative poles and the feedback path $\beta(s)$ is a constant independent of s. For convenience we shall formulate $A(s)$ in the form

$$A(s) = A_0 a_1 a_2 / (s + a_1)(s + a_2) \qquad (6.4.1)$$

where a_1 and a_2 are real and positive and the real quantity A_0 is the value of $A(s)$ for $s = 0$. For the magnitude of the corresponding frequency-response function $A(j\omega)$ we have

$$|A(j\omega)| = A_0 a_1 a_2 / \sqrt{[(\omega^2 + a_1{}^2)(\omega^2 + a_2{}^2)]} \qquad (6.4.2)$$

and this function is constant for $\omega \ll a_1$, a_2 and decreases monotonically with increasing frequency. The frequency-independent feedback path is expressed

$$\beta(s) = \beta_0 \qquad (6.4.3)$$

and from equation (6.2.1) we obtain the following expression for the closed-loop gain $A_C(s)$:

$$\left.\begin{array}{l} A_C(s) = A(s)/[1 - A(s)\beta(s)] \\[2mm] \qquad = A_0 a_1 a_2/[s^2 + (a_1 + a_2)s + a_1 a_2(1 - A_0\beta_0)] \end{array}\right\} \qquad (6.4.4)$$

The poles of $A_C(s)$ are, of course, obtained by factorizing the denominator in (6.4.4) and are at the points

$$\tfrac{1}{2}[-(a_1 + a_2) \pm \sqrt{[(a_1 + a_2)^2 - 4a_1 a_2(1 - A_0\beta_0)]}] \qquad (6.4.5)$$

We now consider four cases which arise as the quantity $(1 - A_0\beta_0)$ increases from negative values towards high positive values.

(a) $(1 - A_0\beta_0) < 0$. The square-root in (6.4.5) exceeds $(a_1 + a_2)$ and one of the poles is real and positive. The system is therefore unstable.

(b) $0 < (1 - A_0\beta_0) < 1$. The poles are real and negative. We see from (6.4.4) that at the lower end of the real-frequency scale $(\omega \ll a_1, a_2)$ the magnitude of $A_C(s)$ is greater than A_0, and accordingly this situation is usually described as one of positive feedback.

(c) $1 < (1 - A_0\beta_0) < (a_1 + a_2)^2/4a_1 a_2$. The poles are still real and negative. Obviously the condition cannot be satisfied for $a_1 = a_2$, but where the open-loop poles are widely separated it describes an important class of negative-feedback systems with non-oscillatory impulse response and (like the open-loop response $A(j\omega)$) having no peak in the magnitude of the frequency response. If we assume that $A_0\beta_0 \gg 1$ we may re-write the condition

$$(1 - A_0\beta_0) < (a_1 + a_2)^2/4a_1 a_2 \qquad (6.4.6)$$

in the approximate form

$$-A_0\beta_0 < a_1/4a_2 \qquad (6.4.7)$$

where a_1 is assumed to be the greater of the two open-loop poles. As the loop gain increases in magnitude towards the limiting value expressed in equations (6.4.6) and (6.4.7), the square-root in (6.4.5) becomes less and the closed-loop poles both approach the point $-\tfrac{1}{2}(a_1 + a_2)$, becoming coincident when the limiting condition is reached. Further increase in $|A_0\beta_0|$ brings us to the fourth region:

(d) $(1-A_0\beta_0)>(a_1+a_2)^2/4a_1a_2$. In this range the quantity under the square-root sign in (6.4.5) becomes negative and the poles are no longer real. As $|A_0\beta_0|$ increases, the poles move away from the real axis, the real part remaining at $-\frac{1}{2}(a_1+a_2)$ and the imaginary part increasing. The impulse response is now oscillatory; we may describe the system in terms of the *LRC* circuit discussed in section 3.4 and say that as the loop gain increases, the equivalent *LC* product falls and the damping ratio ζ also falls. ζ is defined by comparison with equation (3.4.1), by the relation

$$\zeta = (a_1+a_2)/2\sqrt{[a_1a_2(1-A_0\beta_0)]} \qquad (6.4.8)$$

As in section 3.4, the form of the impulse response and of the frequency-response function depend on the value of ζ.

The locus of the closed-loop poles is illustrated in figure 6.3.

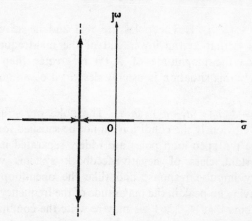

Figure 6.3. Locus of the poles of a second-order feedback system as the loop gain increases.

6.5 Nyquist criterion

We have seen that the closed-loop system function of a feedback system (figure 6.2) is given by the relation

$$A_C(s) = A(s)/[1-A(s)\beta(s)] \qquad (6.5.1)$$

and the system is stable if this function has no poles in the right-half *s*–plane.

An alternative statement of the stability conditions is expressed in the Nyquist criterion.

The feedback system is stable if the locus of the loop-gain $A(j\omega)\beta(j\omega)$ *in the complex plane, plotted out from* $\omega = -\infty$ *to* $\omega = \infty$, *does not enclose the point* $(1+j0)$ *on the real axis* – provided that $A(s)$ and $\beta(s)$ are themselves stable and that $A(s)\beta(s) \rightarrow 0$ as $|s| \rightarrow \infty$.

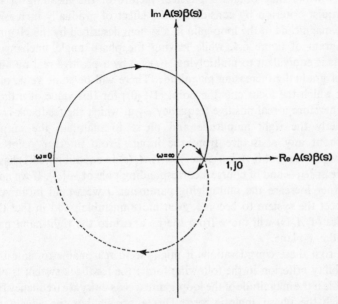

Figure 6.4. A Nyquist diagram. The dotted line shows the region of negative frequency.

An example of a Nyquist diagram is shown in figure 6.4 for a stable feedback system. The three important advantages of this method are:

(a) To construct a Nyquist diagram it is not necessary to know $A(j\omega)\beta(j\omega)$ in algebraic form, provided that experimental results are available for the gain and phase angle in the relevant frequency range.

(b) Even when the algebraic form of $A(j\omega)\beta(j\omega)$ is known, it is often much easier to apply the Nyquist criterion than to go through the procedure of extracting the roots of a polynominal in s – particularly when this is third-order or higher.

(c) The Nyquist method gives a particularly easy way of evaluating the effect of multiplying a given loop-gain function by a constant.

This is of importance in the design of practical feedback systems. For instance, in figure 6.4 we see that the system will become unstable if the magnitude of $A(j\omega)\beta(j\omega)$ is multiplied by about 2, the shape of the locus being unchanged.

Now we may obtain a physical picture of the meaning of the Nyquist criterion by considering the effect of gradually increasing the magnitude of the loop-gain in a system described by the Nyquist diagram of figure 6.4, while leaving the phase angle unchanged. This is equivalent to multiplying $A(s)\beta(s)$ by a positive real number a of gradually increasing magnitude. There will be some value of a for which the locus cuts the point $(1+j0)$; for this value of a there is therefore a real positive frequency ω_1 at which the feedback is of exactly the right amplitude and phase to maintain the output without any assistance from the input. From another point of view, we see that if $A(j\omega_1)\beta(j\omega_1) = 1$, $A_C(s)$ in equation (6.5.1) has a pole at $j\omega_1$ – and of course a corresponding pole at $-j\omega_1$. If we now further increase the multiplying parameter a we would intuitively expect the system to become 'even more unstable', and in fact the poles of $A_C(s)$ will move from the $j\omega$ axis into the right-hand half of the s–plane.

From these considerations it might seem reasonable to state the stability criterion in the following form: the feedback system is unstable if the magnitude of the loop-gain exceeds unity at a frequency for which the phase angle is zero. This approach has the advantage of being 'intuitively obvious' but there are certain cases where it does not give the correct result.

In figure 6.5 there is shown a loop-gain function which is real and greater than $+1$ for two real frequencies, but whose locus does not enclose the $(1+j0)$ point; thus according to the Nyquist criterion the closed-loop system function $A_C(s)$ is stable. A situation of this type is sometimes referred to as one of *Nyquist stability*.

Finally we shall show how the Nyquist criterion is justified. We have seen in section 5.7 that if the locus of a function $F(s)$ is traced out in the $F(s)$ plane as s traverses a contour C in the s–plane, the number of zero-encirclements is equal to the weighted total $\mathcal{N}$ of the poles and zeros of $F(s)$ within the contour C. Now we shall consider the locus of the function $[A(s)\beta(s)-1]$ as s follows a contour along the $j\omega$ axis from $\omega = -\infty$ to $\omega = \infty$, completed by a 'semicircle at infinity' in the right-half s–plane. This is shown in figure 6.6.

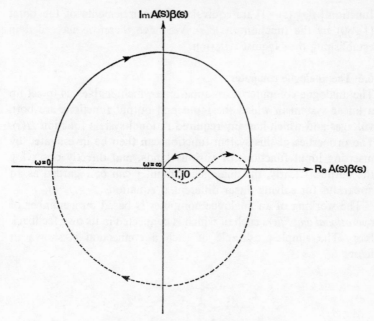

Figure 6.5. Illustration of Nyquist stability.

We are making the following assumptions:

(a) The product $A(s)\beta(s)$ approaches zero as $|s| \to \infty$. This means that the whole region of the 'semicircle at infinity' collapses to a point $(-1+j0)$ in the $[A(s)\beta(s)-1]$ plane, so the locus is essentially from $s = -j\infty$ to $s = j\infty$.

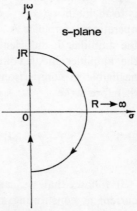

Figure 6.6. Contour in *s*-plane.

(b) The functions $A(s)$ and $\beta(s)$ are both stable, and therefore have no poles within the contour. It follows that $[A(s)\beta(s)-1]$ also has no poles within the contour. The condition $\mathcal{N} = 0$ is therefore equivalent to the condition that $[A(s)\beta(s)-1]$ has no zeros within the contour, that is that $A_c(s)$ is a stable function.

The final step in this argument is to point out that zero-encirclements by the function $[A(s)\beta(s) - 1]$ are equivalent to encirclements of the point $(1 + j0)$ by the function $A(s)\beta(s)$. We have therefore succeeded in establishing the Nyquist criterion.

6.6 The analogue computer

The analogue computer is an apparatus which enables us to set up a linear system in which the input and output functions are both voltages and which has any required rational system function $H(s)$. The properties of this system function can then be investigated by inserting input functions of various kinds, and directly observing the output. Clearly, the analogue computer can be regarded as an apparatus for solving linear differential equations.

The working of an analogue computer is based on a number of *operational amplifiers* each of which is connected in its own feedback loop. The simplest example of such a connection is shown in figure 6.7.

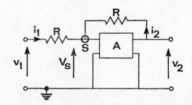

Figure 6.7. Operational amplifier used as an inverter.

This arrangement is designed to act as a voltage inverter, that is to maintain $v_2(t) = -v_1(t)$, $v_2(t)$ being the output voltage of the operational amplifier A. If the system is stable, and if the gain of the amplifier is sufficiently high in the relevant frequency range, the amplifier input voltage v_S and input current i_S will become negligible in comparison with v_1 and v_1/R respectively. We may therefore write

$$i_1 \simeq v_1/R, \ i_2 \simeq v_2/R$$

also $$i_1 + i_2 \simeq 0 \qquad\qquad (6.6.1)$$

so $$v_2 = -v_1.$$

It follows that we can regard the amplifier input voltage and current as constituting an error signal. The point S is called the *summing junction* because it is the terminal at which the feedback current i_2 is added to the input current i_1 to give the error current which constitutes the input to the amplifier A. The approximations that the error voltage and current are both zero are summarized in the statement that the summing junction is a *virtual earth*.

In discussing analogue computers we are always concerned with voltages referred to a common earth terminal, and we use a specialized symbolism in which the earth line is omitted and the amplifier is represented as a triangle (in origin, an arrow pointing in the direction of amplification) with one input and one output lead. In figure 6.8 this symbolism is used to represent a slightly more general form of the circuit in figure 6.7. The two resistors are not equal and in this case we have, in the limit of high amplifier gain, $y(t) = -(R_2/R_1)x(t)$.

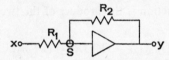

Figure 6.8. An inverter with voltage gain $-(R_2/R_1)$.

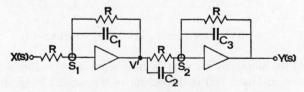

Figure 6.9. An integrator.

Figure 6.9 shows the circuit of an *integrator*. We represent the input and output functions by their transforms $X(s)$ and $Y(s)$; summing currents at S we obtain by the use of the virtual-earth approximation

$$X(s)/R + sCY(s) = 0$$

$$Y(s) = -(1/sRC)X(s) \qquad (6.6.2)$$

The system function of figure 6.10 is of second order.

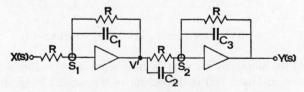

Figure 6.10. Second-order system.

The output V' of the first amplifier is obtained by summing at the first summing junction S_1: we have

$$X(s)/R + V'(s)(sC_1 + 1/R) = 0$$
$$V'(s) = -X(s)/(1 + sRC_1) \tag{6.6.3}$$

and we see that the first amplifier acts as a first-order system with a real pole at $-1/RC_1$. For the second summing junction S_2 the equations are

$$V'(s)(sC_2 + 1/R) + Y(s)(sC_3 + 1/R) = 0$$
$$Y(s) = -V'(s)(1 + sRC_2)/(1 + sRC_3) \tag{6.6.4}$$

showing that the second amplifier acts as another first-order system with a real pole at $-1/RC_3$ and a real zero at $-1/RC_2$. Clearly the overall system function is the product of the two individual system functions:

$$H(s) = Y(s)/X(s) = (1 + sRC_2)/(1 + sRC_1)(1 + sRC_3) \tag{6.6.5}$$

In this case we have built up a second-order system function with real poles from two first-order functions. In principle we could construct a second-order function with complex poles by the use of an inductor as well as a capacitor in the feedback path of an operational amplifier, but inductors are unsatisfactory components in applications of this type and instead we use the three-amplifier system shown in figure 6.11.

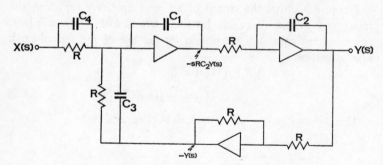

Figure 6.11. Second-order system having complex poles.

Two of the amplifiers are used as integrators and the other is simply an inverter. Since $Y(s)$ is the output of the second integrator, the input of this integrator must be $-sRC_2 Y(s)$, and this quantity is also the output of the first integrator. There are five currents arriving at the first summing junction, and equating the sum of these currents to zero we obtain immediately

$$H(s) = Y(s)/X(s) = (1+sRC_4)/(1+sRC_3+s^2R^2C_1C_2) \qquad (6.6.6)$$

Now this second-order function is a stable one whose poles may be made to take any required real or complex value by appropriate choice of the capacitor values. In practice the circuit of figure 6.11 is only likely to be used when complex poles are required, because real poles are available from the circuit of figure 6.10 using only two amplifiers. Where the required system function has two pairs of complex poles it is sometimes convenient to use two three-amplifier systems in cascade.

By an extension of the methods discussed so far, we can use operational amplifiers, resistors and capacitors to construct any required rational system function, taking suitable precautions to ensure that each individual operational-amplifier loop is stable.

The effect of non-zero error (that is, of non-infinite amplifier gain) can be represented in a very convenient way. Suppose that we are dealing with an amplifier connected to the voltage source through an impedance $Z_1(s)$ and with a feedback impedance $Z_2(s)$. Consider the effect of removing Z_2; the system function will change from approximately $-Z_2(s)/Z_1(s)$ to another function $A'(s)$ which depends to a greater or less extent on Z_1, and which may be called the 'open-loop system function referred to a source with impedance Z_1'. Now since we are interested in the voltage transfer of the system, we may formally represent this situation by the diagram of figure 6.12.

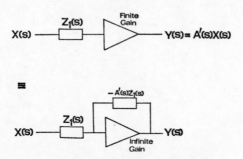

Figure 6.12. Effect of finite amplifier gain.

Here the amplifier is shown as having infinite gain, but having a feedback impedance $-A'(s)Z_1(s)$. The equivalence sign in figure 6.12 is not intended to state that the two representations are equivalent in every way – they have quite different driving-point impedances – but that the second representation is satisfactory when we are considering the voltage transfer. Now it is quite simple to show that when we restore the actual feedback impedance Z_2 to the system, the effect of the finite amplifier gain is correctly represented by the effect of the equivalent impedance $-A'(s)Z_1(s)$ in parallel with $Z_2(s)$, the amplifier being assumed to have infinite gain.

This calculation of the effect of finite error is of particular importance in relation to circuits which are used, not for computation as such, but for the practical construction of system functions of a required type. In recent years it has become increasingly common to use analogue-computer techniques for this purpose – for example it is often cheaper and more practical to construct a resonant circuit by means of transistors in conjunction with resistors and capacitors, rather than by the traditional method of using an inductor and a capacitor.

INDEX

active components 3
all-pass system 46, 55
analogue computer 82
analogue, electrical 6
analytic function 31
amplitude distortion 46
amplitude-modulated carrier 49, 51

band-pass system 48
bilateral Laplace transform 29
buffer 47, 55
Butterworth response 48

carrier 51
Cauchy principal value 63
Cauchy's theorem 32
causality 5, 14, 17, 19, 22, 34, 43, 45, 55, 62
centre frequency 49
closed-loop distortion coefficients 75
closed-loop system function 72
complementary function 20
complex exponential function 18
complex frequency 18
complex impedance 21
conjugate symmetry 22, 25, 50
contour integration 31, 57
control 70, 73
convergence limits 28
convolution operation 16, 30
convolution integral 15, 18, 31, 38
critical damping 40
cut-off frequency 46, 47

damping ratio 39, 51
decay parameter 35
decay time 35
delay, group 45, 53
delay line 45
delay, signal-front 45, 54
delay time 45
delta function 7, 10
differential equation 19
Dirac delta function 7, 10
distortion, non-linear 5, 74
distortion coefficients 74, 75
distortionless system 45, 54
distributed-parameter system 45
driving-point relationship 3, 55, 69
dummy variable 15

emitter-follower 47
envelope function 51
equivalent low-pass system 49
error 70, 71, 73

feedback 70, 71, 72
forced response 21
Fourier coefficients 26, 27
Fourier series 26, 37, 51
Fourier transform 19, 24, 27, 30, 41
Fourier transform of a causal function 43
free response 20
frequency response 14, 22, 24, 41
fundamental frequency 26, 36

87

gain-bandwidth product 61
group delay 45, 53

imaginary-part integral theorem 64
impulse function 7, 10
impulse response 11, 15, 24, 32, 35
initial-value theorem 58
input 1
input terminals 3
integrator 83
intermodulation products 75
inverse Fourier transform 27, 29
inverse Laplace transform 29, 31
inversion integral 31
inverter 83

Kramers-Kronig relations 44, 64

Lag 45, 55
Laplace operator 29, 30
Laplace transform 29, 31, 41
Linear operator 4
Linear-phase system 46
Linearity 3, 5, 74
load impedance 3
logarithmic gain 64
loop gain 72
Lorentz function 50
low-pass system 46
lumped-parameter system 45

magnetic system 2
marginally stable system 35
maximally-flat response 48
mechanical system 2
minimum-phase function 55, 64,
   67, 69
modulation function 51

natural frequency 39
negative feedback 70, 71, 72
negative feedback system 73
non-linear distortion 5, 74

nth order pole 32
Nyquist criterion 78
Nyquist diagram 79
Nyquist stability 80

open-loop distortion coefficients 74
open-loop system function 72
operational amplifier 82
operator, system 1, 4, 16
output 1
output resistance 14
output terminals 3
overshoot 40, 71

partial differential equations 45
partial fractions 34, 58
particular integral 20
passive network 56, 69
phase 25, 44, 54, 64
phase distortion 46
phase lag 45, 55
pole 22, 32
poles and zeros within a contour 69
poles of system function 22, 34, 58,
   67
positive feedback 72

Q 51

rational function 21
reactance integral 64
real-part integral 60, 69
relaxation time 47, 51
required value 70
residue 32, 34, 58, 60
residue theorem 32
response 1
response to sinusoidal input 25
rise time 40

sampling property 10
saturation 5
second-order feedback system 76
second-order pole 35

second-order system 38, 76, 83
servomechanism 70
settling time 40
signal front delay 45, 54
singularity functions 7
stability 4, 14, 19, 22
step function 7, 17
step response 17, 38
summing junction 82
superposition 3
switching transient 21
symmetrical response 49
system function 19, 21, 24, 41
system operator 1, 4, 11, 16, 18

table of transforms 30
temperature-control system 70
thermo-electric system 3
time-constants 12

time-invariance 1
transfer relationship 3, 56
transient 21
transmission line 45

undamped natural frequency 39
unilateral Laplace transform 29
unit impulse function 7, 10
unit step function 7
unstable system 4

virtual earth 83

zero definition 4
zero-encirclements 69, 80
zeros of system function 22, 55, 67